Quality-Driven SystemC Design

Daniel Große · Rolf Drechsler

Quality-Driven SystemC Design

Springer

Dr. Daniel Große
Universität Bremen
AG Rechnerarchitektur
Bibliothekstr. 1
28359 Bremen
Germany
grosse@informatik.uni-bremen.de

Dr. Rolf Drechsler
Universität Bremen
AG Rechnerarchitektur
Bibliothekstr. 1
28359 Bremen
Germany
drechsle@informatik.uni-bremen.de

ISBN 978-90-481-3630-8 e-ISBN 978-90-481-3631-5
DOI 10.1007/978-90-481-3631-5
Springer Dordrecht Heidelberg London New York

Library of Congress Control Number: 2009942231

Cover design: eStudio Calamar S.L.

Printed on acid-free paper

Springer is part of Springer Science+Business Media (www.springer.com)

To Luca, Alia

and

Milena

Contents

List of Figures

List of Tables

Preface

Faced with the steadily increasing complexity and rapidly shortening time-to-market requirements designing electronic systems is a very challenging task. To manage this situation effectively the level of abstraction in modeling has been raised during the past years in the computer aided design community. Meanwhile, for the so-called system-level design the system description language *SystemC* has become the de facto standard. However, while modeling from abstract to synthesizable descriptions in combination with specification concepts like *Transaction Level Modeling* (TLM) leads to very good results, the verification quality is poor. The two main reasons are that (1) the existing SystemC verification techniques do not escort the different abstraction levels effectively and (2) in particular the resulting quality in terms of the covered functionality is only checked manually. Hence, due to the increasing design complexity the number of undetected errors is growing rapidly.

Therefore a *quality-driven design and verification flow* for digital systems is developed and presented in this book. Two major enhancements characterize the new flow: First, dedicated verification techniques are integrated which target the different levels of abstraction. Second, each verification technique is complemented by an approach to measure the achieved verification quality.

The new flow distinguishes three levels of abstraction (namely system level, top level and block level) and can be incorporated in existing approaches. After reviewing the preliminary concepts, in the following chapters the three levels for modeling and verification are considered in detail. At each level the verification quality is measured. In summary, following the new design and verification flow a high overall quality results.

Acknowledgments

We would like to thank the members of the research group for computer architecture at the University of Bremen. A great atmosphere and inspiring discussions are so important.

Furthermore, we would like to thank all co-authors of the papers which formed the starting point for this book: Tim Cassens, Rüdiger Ebendt, Görschwin Fey, Christian Genz, Wolfgang Klingauf, Ulrich Kühne, Hernan Peraza, Robert Siegmund, Tim Warode, and Robert Wille. Especially, we thank Ulrich Kühne and Robert Wille for numerous discussions and successful collaborations.

Many thanks go to Görschwin Fey for proof-reading. We also like to thank Lisa Jungmann for designing the cover page. Cornelia Große improved the readability for non-experts significantly.

Daniel Große and Rolf Drechsler, August 2009

Chapter 1

INTRODUCTION

Over the last decades electronic systems have become more and more important. Nowadays, they play a major role for example in communication, consumer electronic and safety–critical applications. From the design perspective this leads to very diverse requirements. For instance, a design aspect is high computing power as in case of personal computers. For mobile devices like modern cell phones or PDAs rich functionality is the central aspect. In contrast, for safety–critical systems as for example found in automobiles, airplanes or medical devices the design correctness is most important.

Overall the growth of electronic systems is due to the continued advance in the fabrication technology of integrated circuits. As predicted by Gordon Moore in 1965 the number of transistors per chip doubles every 18 months. This exponential growth leads to huge problem instances that have to be handled during the design of a system. For example, Intel has been able to build a processor consisting of 2 billion transistors: Tukwila is scheduled for production in the first quarter of 2010 [Int09].

However, it has been observed that the number of available transistors grows faster than the ability to design chips, which is called *design gap*. But this situation is even intensified if *verification* is considered. Verification refers to the task of ensuring the correct functional behavior of the design and is essential to guarantee high quality systems. Figure 1.1 illustrates both problems [Sem03, Bai04]. As can be seen verification falls behind design and fabrication capabilities which widens the *verification gap*. As a result, several companies meanwhile describe the current situation as *verification crisis* [Sem06]. Thus, in current projects verification engineers outnumber designers, with a ratio of two or three to one [Sem06].

D. Große and R. Drechsler, *Quality-Driven SystemC Design*,
DOI 10.1007/978-90-481-3631-5_1, © Springer Science+Business Media B.V. 2010

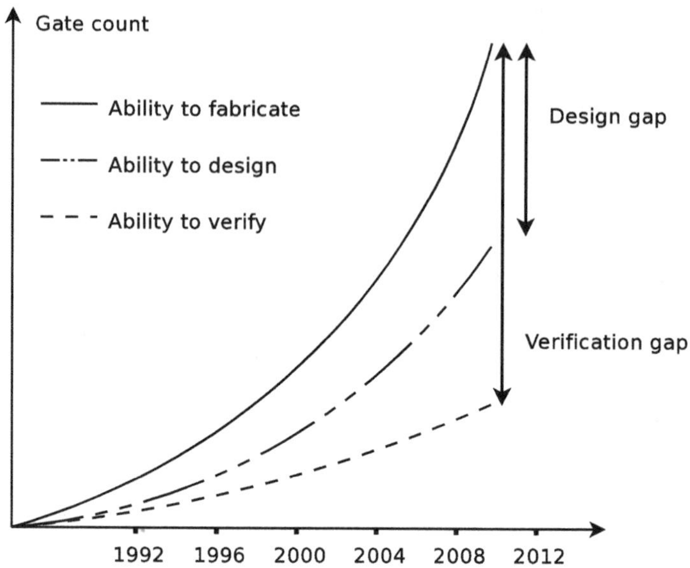

Figure 1.1. Design- and verification gap

In summary, the verification crisis can be attributed to the following points:

- Increasing design sizes.
- Shortening time-to-market demands.
- Besides hardware also software has to be considered.
- Errors are unacceptable in safety–critical systems.
- Rising costs in case of errors.
- *Electronic Design Automation* (EDA) tools are not keeping up with the growing complexity.

To cope with the increasing design sizes the level of abstraction in the design of electronic systems has been raised over the last years. The system engineer writes a C or C++ model of the system to check the concepts and algorithms at the system level. After this design step, the parts of the C/C++ model which have to be implemented in hardware are rewritten in a different language, i.e. typically one of the two major *Hardware Description Languages* (HDLs) Verilog or VHDL is used. This in industry widely applied design flow is depicted in Figure 1.2 where angular boxes denote input/output data and rounded boxes describe a task. However, this design flow has several disadvantages. The most important one is the manual conversion from the

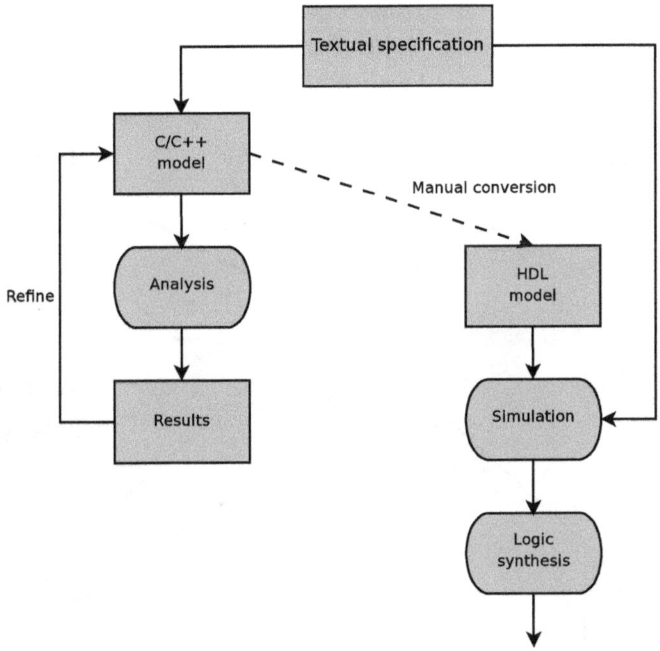

Figure 1.2. System design flow [Sys02]

C/C++ model to the HDL description. Obviously since it is a manual conversion this step is error prone. Furthermore, after conversion the designers only concentrate on the HDL model and hence changes done in the HDL model are not available in the C/C++ model. In addition, for verification a high effort is required since the tests written for the C/C++ model have to be converted to the HDL environment, too. Again, this is a tedious and manual process.

To tackle the decoupling of the system-level model and the HDL description as well as the verification limitations, the C++-based system description language *SystemC* [Sys02, GLMS02] has been developed and standardized by the IEEE [IEE05a]. Using SystemC offers significant advantages over the design flow described above. The proposed SystemC design flow with the subsequent focus on hardware is depicted in Figure 1.3. In the left part of the figure the steps that address the modeling phases of the system are illustrated. Starting from the textual specification the system-level model is written in SystemC. Already at this high level of abstraction the system can be simulated using the SystemC library and a standard C++ compiler to build the *executable specification*. Hence, design space exploration can be performed and hardware/software trade-offs can be checked to meet the requirements of the specification. The parts of the system-level model that have to be implemented in hardware

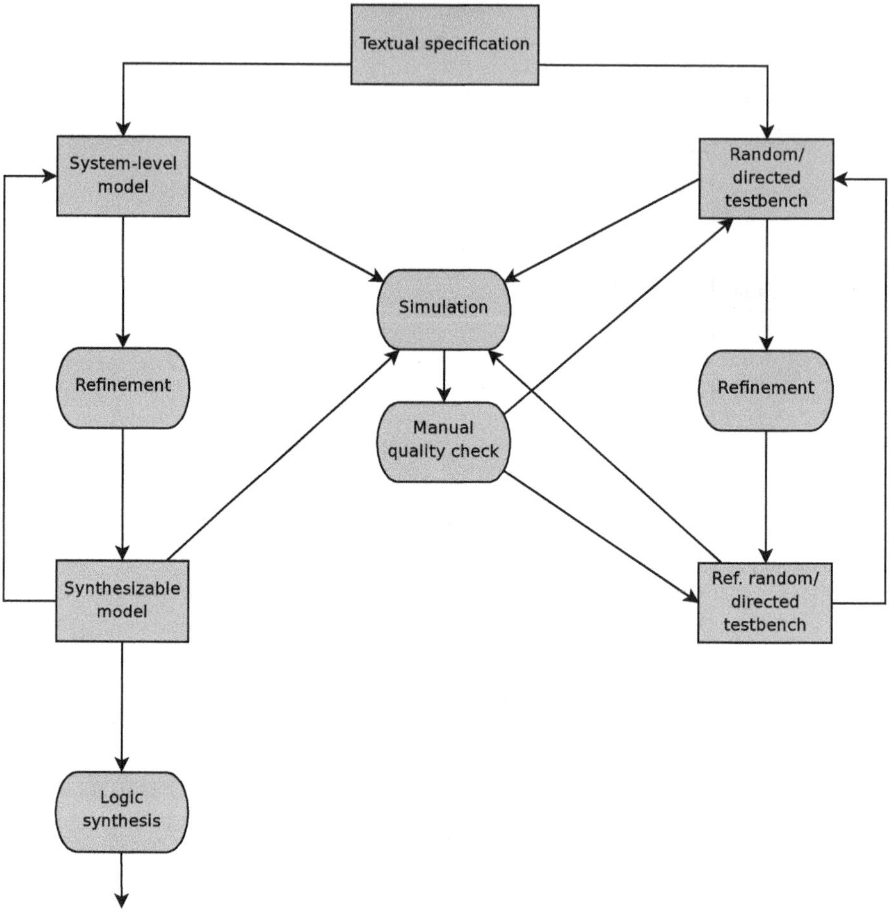

Figure 1.3. SystemC design flow

are not converted into a separate HDL description. In fact, the system is step-wise refined to add hardware and timing constructs until the synthesizable model (typically specified at the *Register Transfer Level* (RTL)) is obtained (see left side of Figure 1.3). Hence, by using SystemC, modeling from the system level to RTL is supported within *one* language. From the verification perspective SystemC allows very fast simulation at different levels of abstraction, i.e. from the abstract descriptions at the system level across the refined models down to the synthesizable model at the RTL. Furthermore, as illustrated in Figure 1.3 on the right, the tests can be reused from the system-level model to the synthesizable model saving conversion time. Reusing the tests, i.e. refining the testbench so that the tests can finally communicate with the synthesizable description gives a higher confidence that the system-level model and the synthesizable model implement the same functionality.

However, the demand for high quality systems reveals the weaknesses of the SystemC design flow. While the modeling from abstract to synthesizable descriptions in combination with specification concepts like *Transaction Level Modeling* (TLM) leads to very good modeling results, the *verification quality* is poor: the SystemC verification techniques do not "escort" the design steps and in particular the resulting quality in terms of the covered functionality is only checked manually. Hence, due to the increasing design complexity the number of undetected errors will grow rapidly and therefore, the verification costs will increase significantly.

In this book a *quality-driven design and verification flow* for digital systems is proposed. Two major enhancements beyond the "traditional" SystemC design flow characterize the new flow: First, dedicated verification techniques are integrated which target the different levels of abstraction. This allows specific improvements of each verification technique and the utilization of formal methods for parts of a verification task or even the entire task. Second, each verification technique is complemented by an approach to measure the achieved verification quality. Hence, a continuous flow results to successfully obtain high quality systems. The flow has been developed at the University of Bremen (see [Gro08]) in collaboration with several industrial partners.

The enhanced design and verification flow is depicted in Figure 1.4. Unchanged data and tasks still have the same color gray as in Figure 1.3, whereas new/modified data and tasks are shown in white. Before the dedicated verification tasks and the respective tasks to check the verification quality – shown in the middle of the figure – are described, we focus on the modeling phases (see left part of Figure 1.4). Like in the "traditional" SystemC design flow the design entry starts with the *system-level model* derived from the textual specification. Then, the system-level model is stepwise refined. In the new flow, the resulting synthesizable model is split up: as a whole the synthesizable model is denoted as *top-level model*, which in turn is divided into several *block-level models*. The reason for this splitting becomes clear in the following when the different verification techniques as well as the complementing quality checks are explained. As depicted in the middle of Figure 1.4 three levels of abstraction for verification and quality checks are distinguished. In the following the approaches developed in this book are described along these three levels of abstraction (a more detailed description is given at the beginning of the respective chapter). Thereby, the presentation follows the verification steps to be performed, but this order is not identical to the three major modeling steps. First, the verification at the system level is described. Due to employed verification techniques and the complementing quality check the verification at the block level is considered next. Finally, verification at the top level is detailed.

System Level. At first, "pure" simulation is replaced by improved constraint-based simulation. The major advantage of this technique is that instead

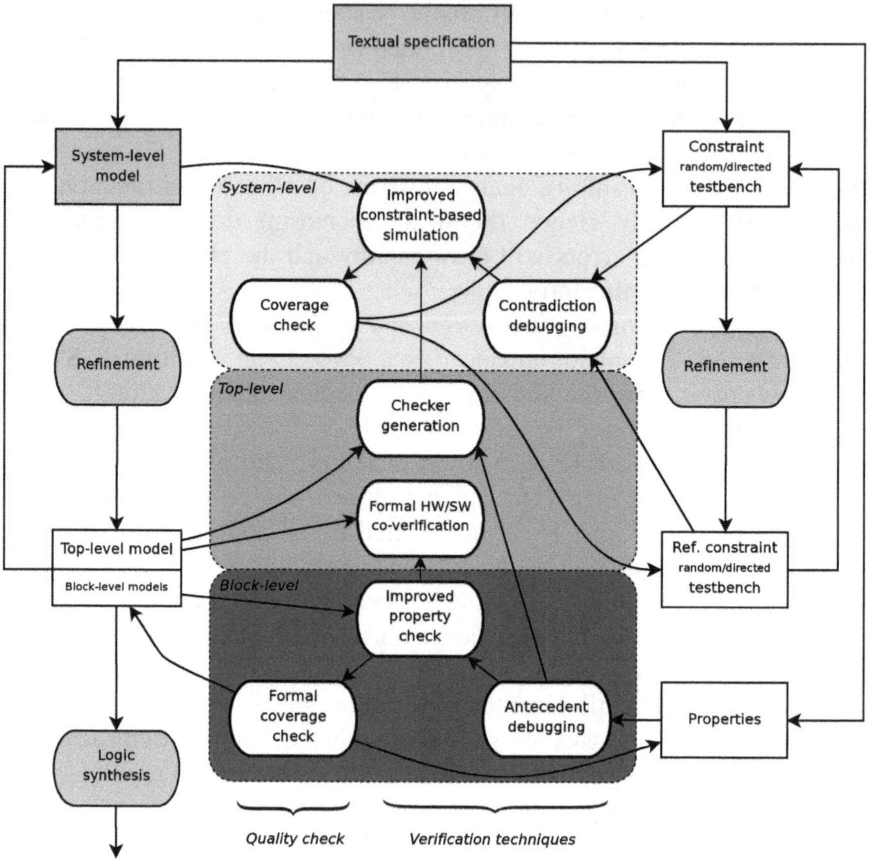

Figure 1.4. Enhanced SystemC design and verification flow

of simulating fixed tests, stimuli are generated that satisfy specified constraints. Thus, scenarios are checked that may be hard to create manually. Particularly corner cases may not be missed which is usually the case in pure simulation. Nevertheless, directed and random tests are still possible. The technology of constraint-based simulation is provided by the *SystemC Verification* (SCV) library. Very good results are obtained by using the already available SCV library. The fundament of the SCV library is a constraint solver that is based on formal methods – *Binary Decision Diagrams* (BDDs) are used to represent the constraints. However, improvements for the SCV library are necessary and have been realized in this work: At first, new operators are added to enrich the constraint specification. Then, the uniform distribution among all solutions of a constraint is guaranteed, which is an important requirement to maximize the chance for entering unexplored regions of the design state space.

Besides the core aspects of constraint-based simulation within the SCV library, during the formulation of complex non-trivial constraints over-constraining occurs, i.e. there is no solution for all constraints. In practice, this is often the case if constraints are added to drive the simulation into a certain direction. To handle this problem, a new contradiction debugging method is introduced that automatically identifies all contradictory constraint expressions and thus the manual debugging time is reduced. In Figure 1.4 the approach is denoted as *contradiction debugging*.

To check the resulting verification quality an approach is presented to measure "how thorough" the design was tested (denoted as *coverage check* in the figure). The method is based on dedicated code coverage techniques that have been developed for SystemC models. A coverage report is generated that presents all parts of the model that have not been executed during simulation. Thereby, feedback about the achieved verification quality is provided and the manual quality check is removed.

Block Level. Continuing the design process from the system-level model (see left part of Figure 1.4), the proposed flow separates the synthesizable model into the top-level model and several block-level models. This splitting results from the fact that different verification techniques are used at both levels. For the block-level models, formal methods are applied that allow to guarantee the design correctness (in general, at the top-level this is not possible, since the resulting models become too complex). Hence, before investing much effort in the verification at the top level, block-level verification is performed.

Therefore, a property checking method for SystemC is presented. For property checking the temporal properties are specified in the standardized *Property Specification Language* (PSL). Furthermore, all properties have the form of an implication and are defined over a bounded time interval. Thus, based on *Bounded Model Checking* (BMC) the properties are proven if the corresponding *Boolean Satisfiability* (SAT) instance is unsatisfiable. In addition, the iterative application of property checking is improved as follows. Usually, at the beginning the verification engineer specifies a property with a "strong" antecedent. Then, this antecedent is stepwise weakened and the property check is performed again. For this scenario a speed-up of the underlying SAT proof is achieved by reusing learned information from a previous run. Property checking as well as the enhancement are shown as the combined task *improved property check* in Figure 1.4.

Typically, in the antecedent of the property the assumptions about the design environment are specified and joined by logical AND. However, for complex designs and hence non-trivial properties the verification engineer may be confronted with the problem of an overall conjunction which has no solution, i.e. the antecedent is contradictory. Since in this case a property trivially holds, this situation has to be avoided. An automatic approach is presented for debugging of a contradictory antecedent. The basic principles are similar to the contradiction debugging at the system level. However, for property checking the approach has to distinguish whether a contradiction results from the antecedent solely or from the antecedent *and* parts of the design (see task *antecedent debugging* in Figure 1.4).

Again, at this level the quality of verification is ensured. Therefore, an approach to analyze coverage in property checking is presented. With this technique the completeness of a property set can be shown, i.e. there is no scenario where the behavior of the design is not determined by the properties. If such a scenario – a gap – is found, this gap is presented to the user. How to deal with uncovered scenarios to achieve full coverage is discussed in detail. In Figure 1.4 the corresponding task is *formal coverage check.*

Top level: To bridge the verification from block level to system level checkers are generated. The motivation for this procedure is that the verified block-level parts exchange data using complex communication mechanisms and hence this is the focus of verification here. The basic idea of checkers is to embed temporal properties after a respective transformation directly into the SystemC description. The checkers can also be reused from the block-level verification. They are validated at the top level during simulation. Hence, the system level coverage check for ensuring the verification quality is also available here. Besides simulation, the checkers are also synthesizable and can be utilized for on-line tests after fabrication. The task *checker generation* denotes this method in Figure 1.4.

Finally, a formal hardware/software co-verification approach for embedded systems is introduced. The approach is based on BMC and the formal coverage check. At first, the correctness of the underlying hardware is proved. Then, the hardware/software interface is considered and verified applying BMC. Based on this result programs using the hardware/software interface can be formally verified. This task is depicted in Figure 1.4 as *formal HW/SW co-verification.*

All mentioned approaches have been implemented and evaluated in several experiments. In summary, the main advantages of this advanced and innovative design and verification flow are:

- System design and verification flow covering all levels of abstraction
- Dedicated verification techniques at each level of abstraction
- System-level verification by improved constraint-based simulation

- Block-level verification based on improved property checking
- Top-level verification with checkers and formal hardware/software co-verification
- Integration of debugging approaches to identify contradictions in specified tests
- High verification quality due to automatic coverage checks
- Usage of code coverage techniques to ensure system-level and top-level verification quality
- Formal coverage analysis for property checking to guarantee verification quality at the block level

In the following chapters of this book the approaches themselves are described, and related work, experimental results and directions for further research are given.

This book is structured as follows: Chapter 2 gives the basic notations, the background on formal verification and a brief introduction to SystemC. Chapter 3 presents the system-level verification methods. Improvements for constraint-based simulation in the SCV library, the approach for constraint contradiction debugging, and the method to guarantee the verification quality of the testbench using code coverage are detailed. In Chapter 4 block-level verification is considered. First, the SystemC property checker is introduced. Then, the improvement to accelerate the iterative application of property checking is described. The next section provides the method for automatic antecedent debugging. Afterwards, the formal coverage analysis approach to ensure the resulting verification quality is presented. Chapter 5 describes the method for checker generation. Moreover, the formal hardware/software co-verification method for embedded systems is introduced. Finally, in Chapter 6 the book ends with a summary and conclusions.

Chapter 2

PRELIMINARIES

In this chapter the basic notations and definitions are given to make this book self-contained. In the first part Boolean reasoning including state-of-the-art proof techniques are reviewed. Then, circuits and their respective representation are introduced. Thereafter, the two typical scenarios of formal verification are described, i.e. equivalence checking and (bounded) model checking. Finally, the basics and concept of the system description language SystemC are given.

The presentation is always given in a compact way, but references for further reading are provided.

2.1 Boolean Reasoning

First, in this section the basics regarding Boolean functions and the important mathematical structure Boolean algebra are given. Then, binary decision diagrams and Boolean satisfiability are reviewed. For more details we refer the reader, e.g., to [Weg87, HS96, DB98, GG07].

2.1.1 Basic Notations and Boolean Algebra

Boolean variables may assume values from the set $\mathbb{B} := \{0, 1\}$.

DEFINITION 2.1 *A* Boolean function *f is a mapping from $f : \mathbb{B}^n \rightarrow \mathbb{B}$, $n \in \mathbb{N}$. Usually, f is defined over the finite set of Boolean variables $X_n := \{x_1, x_2, \ldots, x_n\}$ and hence is denoted by $f(x_1, \ldots, x_n)$. A* multi-output Boolean function *f is a mapping from $f : \mathbb{B}^n \rightarrow \mathbb{B}^m$ with $n, m \in \mathbb{N}$ and $m \geq 2$.*

Since the Boolean algebra is the basis for digital circuits its definition is given in the following.

D. Große and R. Drechsler, *Quality-Driven SystemC Design*,
DOI 10.1007/978-90-481-3631-5_2, © Springer Science+Business Media B.V. 2010

DEFINITION 2.2 *A* Boolean algebra *is a set A with two binary operations* $+$ *and* $\cdot$, *one unary operation* $^{-}$ *and two distinct elements* 0 *and* 1 *such that for all elements* $x_1, x_2, x_3 \in A$ *the following holds:*

Associative laws:	$x_1 + (x_2 + x_3) = (x_1 + x_2) + x_3$
	$x_1 \cdot (x_2 \cdot x_3) = (x_1 \cdot x_2) \cdot x_3$
Commutative laws:	$x_1 + x_2 = x_2 + x_1$
	$x_1 \cdot x_2 = x_2 \cdot x_1$
Distributive laws:	$x_1 \cdot (x_2 + x_3) = (x_1 \cdot x_2) + (x_1 \cdot x_3)$
	$x_1 + (x_2 \cdot x_3) = (x_1 + x_2) \cdot (x_1 + x_3)$
Complements:	$x_1 + \overline{x}_1 = 1$
	$x_1 \cdot \overline{x}_1 = 0$
Absorption:	$x_1 + (x_1 \cdot x_2) = x_1$
	$x_1 \cdot (x_1 + x_2) = x_1$

EXAMPLE 2.3 *If we choose* $A = \mathbb{B}$ *and define the operations* $+$, $\cdot$ *and* $^{-}$ *as shown in Table 2.1, then the Boolean algebra results that is used for describing the behavior of circuits.* $+$ *is called* disjunction *or* OR-function, $\cdot$ *is called* conjunction *or* AND-function, *and* $^{-}$ *is called* negation *or* NOT-function, *respectively. Sometimes, instead of the symbols* $+, \cdot, ^{-}$ *the symbols* $\vee, \wedge, \neg$ *are used, respectively.*

From Boolean variables $x_i \in X_n$, constants 0, 1, and the operations $+$, $\cdot$, $^{-}$ and parentheses $(,)$ *Boolean expressions* can be formed. Note that every Boolean function can be written as a Boolean expression. In addition to the already introduced operations, the following operations are important:

- XOR: $x_1 \oplus x_2 := x_1 \cdot \overline{x}_2 + \overline{x}_1 \cdot x_2$
- Implication: $x_1 \rightarrow x_2 := \overline{x}_1 + x_2$
- Equivalence: $x_1 \leftrightarrow x_2 := \overline{x_1 \oplus x_2}$

Finally, the *positive and negative cofactors* of a Boolean function $f : \mathbb{B}^n \rightarrow \mathbb{B}$ with respect to a variable x_i are $f_{x_i} := f(x_1, \ldots, x_{i-1}, 1, x_{i+1}, \ldots, x_n)$ and $f_{\overline{x}_i} := f(x_1, \ldots, x_{i-1}, 0, x_{i+1}, \ldots, x_n)$, respectively.

Table 2.1. Operations for Boolean algebra of circuits

x_1	x_2	$x_1 + x_2$
0	0	0
0	1	1
1	0	1
1	1	1

x_1	x_2	$x_1 \cdot x_2$
0	0	0
0	1	0
1	0	0
1	1	1

x_1	$\overline{x}_1$
0	1
1	0

2.1.2 Binary Decision Diagrams

A Boolean function $f : \mathbb{B}^n \rightarrow \mathbb{B}$ can be represented by a *Binary Decision Diagram* (BDD) [Bry86] which is a directed acyclic graph where a *Shannon decomposition*

$$f = \overline{x}_i f_{\overline{x}_i} + x_i f_{x_i} \quad (1 \leq i \leq n)$$

is carried out in each node. The function which is represented by an internal node of a BDD is determined recursively by the two children, whereas the terminal nodes represent the two constant functions 0 and 1. With respect to the carried out Shannon decomposition of a node, f_{x_i} is called the *high child* and $f_{\overline{x}_i}$ is called the *low child*, respectively.

A BDD is called *ordered* if each variable is encountered at most once on each path from the root to a terminal node and if the variables are encountered in the same order on all such paths. Formally, the resulting *variable ordering* is nothing else than a mapping $o : \{1, \ldots, n\} \rightarrow X_n$, where $o(i)$ denotes the ith variable in the ordering.

A BDD is called *reduced* if it does neither contain isomorphic subgraphs nor does it have redundant nodes. Reduced and ordered BDDs (ROBDDs) are canonical representations, i.e. the BDD representation for a given Boolean function is unique as long as the variable ordering is fixed [Bry86]. In the following, we refer to reduced and ordered BDDs for brevity as BDDs. The size of a BDD is given by the number of non-terminal nodes.

It is well known that BDDs are very sensitive to the chosen variable ordering, i.e. the size may vary from linear to exponential.

EXAMPLE 2.4 *Consider the BDDs for the function* $f = x_1x_2 + x_3x_4 + \ldots + x_{2n-1}x_{2n}$ *with* $n = 3$ *depicted in Figure 2.1. Dashed lines are used for the 0-assignment, while solid lines denote the 1-assignment. For the ordering* $o_1 = (x_1, x_2, \ldots, x_{2n-1}, x_{2n})$ *the BDD shown in Figure 2.1(a) has a size of* $O(n)$*, whereas the BDD shown in Figure 2.1(b) with the ordering* $o_2 = (x_1, x_3, \ldots, x_{2n-1}, x_2, x_4, \ldots, x_{2n})$ *has a size of* $O(2^n)$.

Since the problem to decide whether a given variable ordering can be improved is NP-complete [BW96], several heuristics to find a good variable ordering have been proposed. There are topology-based heuristics using structural information [FOH93] and dynamic reordering techniques like *sifting* [Rud93] which are based on level exchanges. For scalable design descriptions word-level information can be used to learn a good ordering for the small instance and extrapolate this for the large instance [GD03a]. Furthermore, methods using evolutionary algorithms have been proposed [DBG96] which can obtain better results than the approaches mentioned before but usually the run-time is much higher.

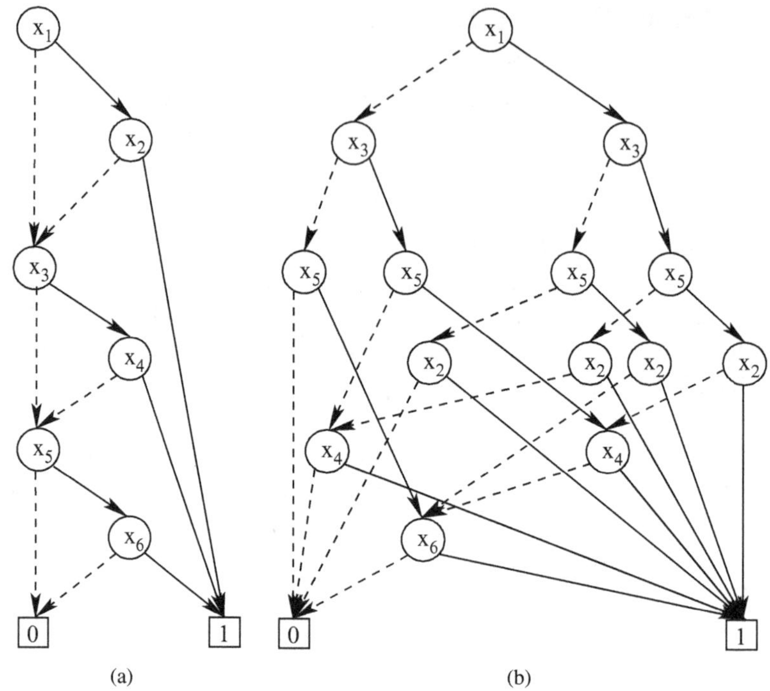

Figure 2.1. BDDs for function $f = x_1x_2 + x_3x_4 + x_5x_6$

To further minimize the size of the representation *complement edges* have been introduced [BRB90]. They allow to represent both a function and its complement by the same node, modifying the edge pointing to the node instead. Besides a compact representation of Boolean functions efficient techniques for symbolic manipulations of BDDs exist. The core operation used in the recursive manipulation algorithms is the *If-Then-Else* (ITE) operator:

$$ite(f, g, h) = f \cdot g + \overline{f} \cdot h$$

With the recursive formulation

$$ite(f, g, h) = \overline{x}_i \cdot ite(f_{\overline{x}_i}, g_{\overline{x}_i}, h_{\overline{x}_i}) + x_i \cdot ite(f_{x_i}, g_{x_i}, h_{x_i})$$

the computation of an operation is determined. As a result the binary operations $+$ and $\cdot$ can be implemented efficiently as graph algorithms, i.e. given two BDDs of size $|F|$ and $|G|$ the complexity for the operations is $O(|F| \cdot |G|)$. Furthermore, *existential quantification* $\exists x_i f = f_{\overline{x}_i} + f_{x_i}$ as well as the *universal quantification* $\forall x_i f = f_{\overline{x}_i} \cdot f_{x_i}$ can be implemented such that the complexity is polynomial in the BDD sizes.

For the practical application of BDDs very efficient implementations have been proposed. A widely used high quality BDD package is the *Colorado University Decision Diagram Package* (CUDD) [Som01]. Basically, in such a package a BDD node is represented as a triple `(index(v), high(v), low(v))`, where the first entry gives the index of the variable to which v is related, and `high(v)` and `low(v)` are pointers to the memory locations of the high and low child of v, respectively. Complement edges are realized by using the least significant bit of these pointers to store the information about complementation. This is feasible since the memory is addressed word-wise and hence the least significant bit of a pointer is always zero. All the node triples are stored in a hash table (the so-called *unique table*). Then, whenever an operation needs to add a new BDD node it is checked if the node already exists in the table. By this, a BDD is automatically build in the reduced form. For the above mentioned symbolic manipulation operations a second hash table, the *computed table*, is employed. As hash key the two pointers of an operation and the operation itself are used and as value the resulting node is stored. Since during the computation of operations each result is stored in the computed table, repeated computations are avoided.

BDDs can also be used to represent multi-output functions. Then, a BDD for each component function is used in the *shared BDD* representation. Note that in this case the variable ordering is the same for each BDD.

2.1.3 Boolean Satisfiability

Studies have shown that *Boolean satisfiability* (SAT) and BDDs are orthogonal techniques for Boolean reasoning. From a theoretical point of view this can be proven [GZ03]. In particular, SAT has shown advantages in formal verification where often a single counter-example is sufficient. Hence, SAT addresses the scalability and performance limitations of BDD-based methods. In the following the definition of the SAT problem as well as an example and basic concepts to solve an instance of SAT are provided.

DEFINITION 2.5 *The* Boolean Satisfiability *(SAT) problem is to determine an assignment* $\alpha = (a_1, \ldots, a_n)$ *to variables of a Boolean function* $f(x_1, \ldots, x_n)$ *such that* $f(\alpha) = 1$ *(i.e.* f *is satisfiable) or to prove that no such assignment exists (i.e.* f *is unsatisfiable).*

SAT is one of the central NP-complete problems. In fact, it was the first known NP-complete problem that was proven by Cook in 1971 [Coo71]. Despite the complexity of the SAT problem, in practice many large SAT instances can be solved today.

Usually, a SAT instance is represented as a Boolean formula in *Conjunctive Normal Form* (CNF) which is given as a set of clauses; each clause is a set of literals and each literal is a propositional variable or its negation. The CNF

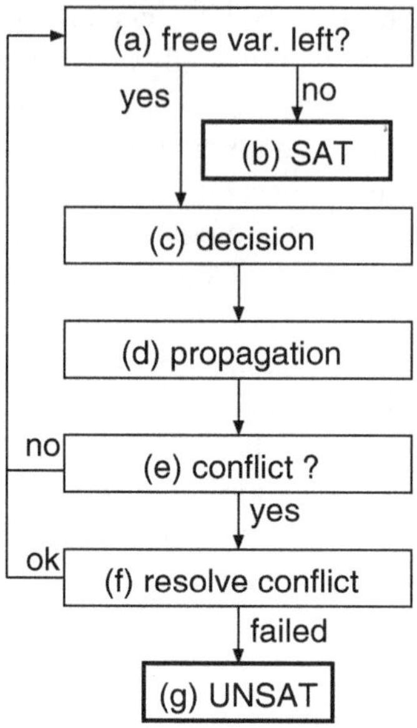

Figure 2.2. DPLL algorithm in modern SAT solvers

formula is satisfied if all clauses are satisfied. A clause is satisfied if at least one of its literals is satisfied. The positive literal x is satisfied if 1 is assigned to the variable x. The negative literal $\overline{x}$ is satisfied if 0 is assigned to the variable x.

EXAMPLE 2.6 *The following Boolean formula f is given in CNF:*

$$f = (x_1 + x_2 + \overline{x}_3) \cdot (\overline{x}_1 + x_3) \cdot (\overline{x}_2 + x_3)$$

Then $x_1 = 1, x_2 = 1$ and $x_3 = 1$ is a satisfying assignment for f. The values of x_1 and x_2 ensure that the first clause becomes satisfied, while x_3 ensures this for the remaining two clauses.

The basic search procedure to find a satisfying assignment is shown in Figure 2.2 and follows the structure of the DPLL algorithm [DP60, DLL62]. Instead of simply traversing the complete space of assignments, intelligent decision heuristics and other improvements that are explained below lead to an effective search procedure. The description follows the implementation of the procedure in modern SAT solvers. While there are free variables left (a), a decision is made (c) to assign a value to one of these variables. Then, implications are determined due to the last assignment by *Boolean Constraint Propagation* (BCP) (d). This may cause a conflict (e) that is analyzed. If the conflict can be resolved by undoing assignments from previous decisions, backtracking is done (f). Otherwise, the instance is unsatisfiable (g). If no further decision can be done, i.e. a value is assigned to all variables and this assignment did not cause a conflict, the CNF is satisfied (b). The *decision level* denotes the number of variables assigned by decisions in the current partial assignment, i.e. neglecting variable assignments due to implications.

During the last years several techniques have been developed to improve the efficiency of SAT solvers. They can be categorized into conflict based learning with non-chronological backtracking, intelligent decision heuristics, and improvements of BCP. A brief discussion is given in the following.

Conflict based learning with non-chronological backtracking was proposed in [MSS96]. The main idea is that a detected conflict may have been caused by much earlier assignments and hence it is possible to jump back to an earlier decision level than in the original SAT algorithms. For conflict based learning usually all implications are stored in an implication graph which describes the implication relationships of variable assignments. By forming a cut on the implication graph, i.e. to identify a set of assignments that are sufficient to cause a conflict a so-called *conflict clause* is build and added to the SAT instance. This clause prevents the SAT solver to reenter the same non-solution space again and is also used to determine the respective decision level to backtrack to. Of course the conflict clauses have to be deleted from time to time since otherwise they might cause a memory problem. Details how this can be achieved can be found in [MS99, ZMMM01, ES04].

Another major improvement results from sophisticated decision heuristics. Several decision heuristics have been proposed. Very often statistical data like the occurrence of literals in clauses forms the basis for the heuristics. For example, a heuristic which emphasizes the role of recently learned conflict clauses is the *Variable State Independent Decaying Sum* (VSIDS) [MMZ$^+$01]. This heuristic uses counters for each literal. They are updated accordingly each time a new conflict is learned and all the counters are divided by a constant periodically. As a result, the literals of recently learned clauses are preferred for the upcoming decisions.

The efficiency of BCP as described above is very important for the overall performance of a SAT solver. After each decision BCP is called. A very efficient implementation is the *two literal watching scheme* found in the SAT solver *Chaff* [MMZ$^+$01]. The basic idea is to watch two literals per clause which are used to detect whether a conflict or an implication is possible since assigning a literal to false not necessarily causes a conflict or implication. In total, during the search not all the time each clause has to evaluated. This results in a significant speed-up.

Because of all the described advances, SAT solvers are widely used in computer-aided design today. For example, SAT is applied in domains like verification [BCCZ99, KPKG02], automatic test pattern generation [Lar92, DEF$^+$08], diagnosis [SVV04, FSBD08], and synthesis [ZSM$^+$05, GWDD09].

2.2 Circuits

First, in this section the formal model of sequential circuits is introduced and a simple example is provided. Then, in the second part the transformation of basic circuit elements into CNF is given.

2.2.1 Modeling of Sequential Circuits

A sequential circuit is composed of memory elements (like flip-flops or registers) and combinational logic. The combinational logic consists of elements from a set of basic gates (called *gate library*). A typical library is shown in Figure 2.3. It contains the standard gates NOT, AND, OR and XOR. In the figure also the respective Boolean functions are given.

Usually, the behavior of a sequential circuit is modeled as a *Finite State Machine* (FSM) where the output is associated to each state transition. Such a FSM is called a *Mealy machine* and is defined as

DEFINITION 2.7 *A* Mealy machine *is a 6-tuple* $M = (I, O, S, S_0, \delta, \lambda)$, *where*

- I *is the finite set of inputs,*
- O *is the finite set of outputs,*
- S *is the finite set of states,*
- $S_0 \subseteq S$ *is the set of initial states,*

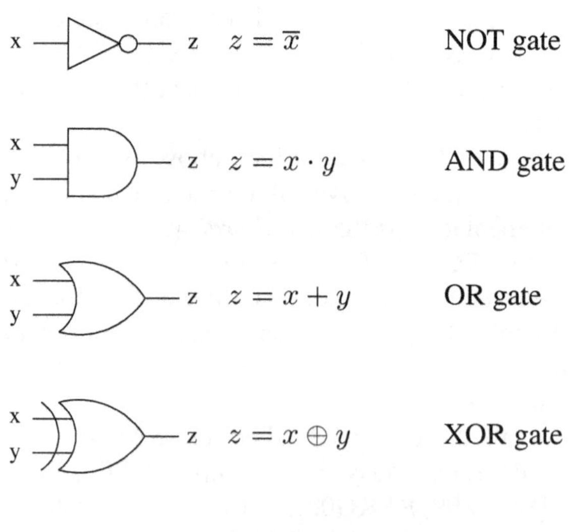

Figure 2.3. Library of basic gates

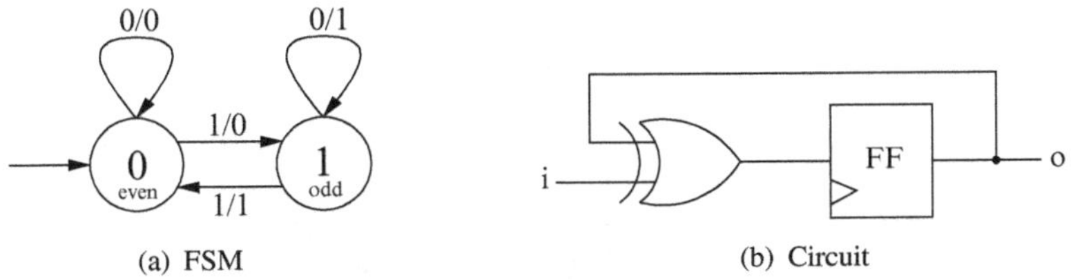

Figure 2.4. Odd parity checker

- $\delta : I \times S \to S$ *is the state transition function, and*
- $\lambda : I \times S \to O$ *is the output function.*

Since we consider circuits, the set of inputs, outputs and states uses a binary encoding, i.e. the corresponding sets are defined over the Boolean values $\mathbb{B}$ as: $S := \mathbb{B}^n, I := \mathbb{B}^m$ and $O := \mathbb{B}^k$.

In the following a simple example of an FSM and the corresponding sequential circuit is given.

EXAMPLE 2.8 *Consider the FSM in Figure 2.4(a). This FSM realizes an odd parity checker, i.e. the output becomes one whenever the processed input data has an odd number of 1's. In the depicted diagram the states are represented as nodes, whereas the transitions between the states are given as arrows. The arrows are labeled with the corresponding input followed by the output. If for this FSM the state is denoted by variable s, the input by variable i and the output by variable o, then the state transition function is $\delta(i, s) = s \oplus i$ and the output function for o is $\lambda(i, s) = s$. The corresponding sequential circuit is depicted in Figure 2.4(b). It has one input i and one output o. The internal output of the XOR gate is connected to a flip-flop denoted as* `FF` *in the figure. We assume that the initial state of this flip-flop is 0 (which cannot be seen in the figure) since the initial state of the FSM is the state* `even`.

2.2.2 CNF Transformation

As described in Section 2.1.3 SAT is applied for solving different problems in computer-aided design. However, to apply SAT an efficient translation of a circuit into CNF is necessary. The principle transformation in the context of Boolean formulas has been proposed by Tseitin [Tse68]. The Tseitin transformation can be done in linear time maintaining satisfiability. This is achieved by introducing a new variable for each sub-formula and constraining that this new variable is equivalent to the sub-formula. For circuits the respective transformation has been presented in [Lar92]. We only briefly summarize the results of the transformation in Table 2.2. Note that by applying the transformation

Table 2.2. CNF for basic gates

Gate	Boolean function	CNF
NOT	$z = \overline{x}$	$(x + z) \cdot (\overline{x} + \overline{z})$
AND	$z = x \cdot y$	$(\overline{z} + x) \cdot (\overline{z} + y) \cdot (z + \overline{x} + \overline{y})$
OR	$z = x + y$	$(z + \overline{x}) \cdot (z + \overline{y}) \cdot (\overline{z} + x + y)$
XOR	$z = x \oplus y$	$(\overline{z} + x + y) \cdot (z + \overline{x} + y) \cdot (z + x + \overline{y}) \cdot (\overline{z} + \overline{x} + \overline{y})$

redundant clauses might result. But they can be eliminated using resolution. The table shows the minimized CNF already. The CNF for the complete circuit is formed by the conjunction of all "local" CNFs. The presented transformation is used very often. Of course also optimizations have been proposed like for example to merge gates and hence reduce the number of auxiliary variables [Vel04].

2.3 Formal Verification

The main idea of formal verification is to prove the functional correctness of a system. The two typical scenarios – equivalence checking and model checking – that build the basis of formal verification are described in the following. For a more detailed presentation we refer the reader to [Kro99, CGP99, Dre04, GG07].

2.3.1 Equivalence Checking

The task of *equivalence checking* is to formally prove that two circuit representations are functional equivalent. In the following we only briefly review equivalence checking of combinational circuits. Hence, we assume that both circuits have the same number of states and that a matching between them is known. Then, equivalence checking can be carried out by building the so-called *miter circuit* [Bra93] (see Figure 2.5). First, the corresponding inputs of both circuits are connected. Then, each corresponding output pair is fed into an XOR, whereas all XORs are fed into a large OR which becomes the overall output of the miter structure. This output is 1 if there exists an input assignment to the miter circuit which causes at least one output pair producing different values.

Obviously, for solving the equivalence checking problem BDDs and SAT techniques can be used. However, for large circuits the BDD-based approaches suffer from memory explosion. Hence, SAT-based approaches have been considered (see, e.g., [GPB01, DS07]) where the (partial) miter is translated to CNF and the overall output constrained to 1. A combination of proof engines which exploits structural information and merges identical circuit parts has been considered in [KPKG02].

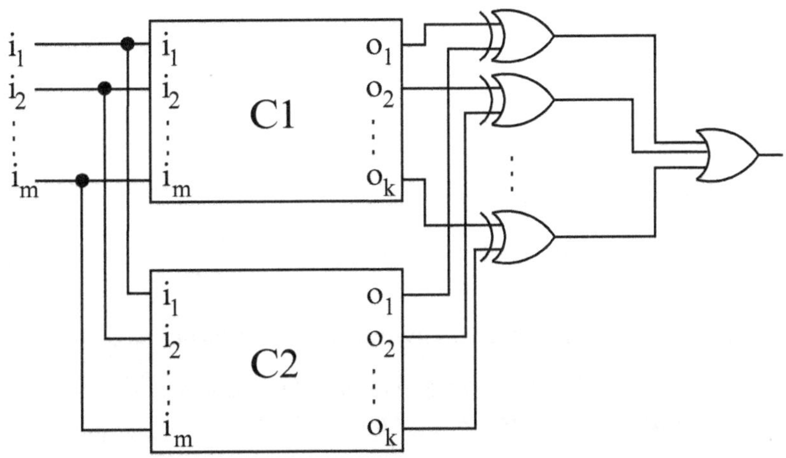

Figure 2.5. Miter circuit

The presentation of equivalence checking is mainly given here to introduce both typical scenarios of formal verification. Of course in the literature also the more general problem of sequential equivalence checking is considered (for an overview see [MS05]). Besides this, also first approaches to show equivalence between high-level descriptions, like for example C descriptions and circuit implementations, have been proposed. But in this context, the research is still at the very beginning (see, e.g., [SMS$^+$07] where as high-level language SpecC is used).

2.3.2 Model Checking

The idea of model checking is to verify whether a model – derived from hardware or software – satisfies a formal specification. This specification is described by temporal logic formulas such as *Linear Time Logic* (LTL) [Pnu77] formulas or *Computation Tree Logic* (CTL) [BAMP81] formulas. The model represents the behavior of the considered system. Usually, the temporal logic specification is interpreted in terms of a *Kripke structure.*[1] Hence, first Kripke structures are introduced. Then, the transformation of a Mealy machine into a Kripke structure is described. On top of these models the temporal logic for specification of properties is presented. Finally, algorithms for model checking are reviewed.

[1] A Kripke structure is also called temporal structure.

Models

DEFINITION 2.9 *A* Kripke structure *K is a 5-tuple $K = (S, S_0, R, V, L)$, where*

- *S is the finite set of states,*
- *$S_0 \subseteq S$ is the set of initial states,*
- *$R \subseteq S \times S$ is the transition relation with $\forall s \in S\ \exists s' \in S : (s, s') \in R$ (i.e. R is total),*
- *V is the set of propositional variables (or atomic formulas), and*
- *$L : S \to \mathcal{P}(V)$ is the labeling function that labels each state with the subset of variables that are true in that state.*

A Kripke structure determines the set of computations of a system as an "unrolled" tree structure. For a sequential circuit given as a Mealy machine according to Definition 2.7 the corresponding Kripke structure is constructed as follows:

DEFINITION 2.10 *Let $M = (I_M, O_M, S_M, S_{0_M}, \delta, \lambda)$ be a Mealy machine. Then, the Kripke structure $K = (S, S_0, R, V, L)$ for M is defined as*

- *$S := I_M \times S_M$,*
- *$S_0 := \{(i, s) | i \in I_M, s \in S_{0_M}\}$,*
- *$R \subseteq S \times S$ with $((i, s), (i', s')) \in R$ if $s' = \delta(i, s)$,*
- *$V := \{\widehat{i_1}, \widehat{i_2}, \ldots, \widehat{i_m}\} \cup \{\widehat{o_1}, \widehat{o_2}, \ldots, \widehat{o_k}\} \cup \{\widehat{s_1}, \widehat{s_2}, \ldots, \widehat{s_n}\}$ is the set of new variables for the inputs, outputs and states of M, and*
- *$L((i, s)) := \bigcup_{j=1}^{m} \{\widehat{i_j} \mid i_j = 1\} \cup \bigcup_{j=1}^{k} \{\widehat{o_j} \mid \lambda_j(i, s) = 1\} \cup \bigcup_{j=1}^{n} \{\widehat{s_j} \mid s_j = 1\}$.*

A state s' is the next state of s in the Kripke structure K if there exist values for the inputs of M such that M makes a transition from s to s'. All possible inputs for the Mealy machine are encoded in the states of the Kripke structure. The labeling function is formed such that each state is marked with the input, output and state variables that are true in that state. In the following definition a sequence of states (also called *path*) is considered since the semantics of temporal logic formulas are interpreted over paths.

DEFINITION 2.11 *Let $K = (S, S_0, R, V, L)$ be a Kripke structure. A* path *π is a infinite sequence of states, i.e. $\pi = \langle s^0, s^1, \ldots \rangle$ with $\forall j \geq 0 : (s^j, s^{j+1}) \in R$. A suffix of a path $\pi = \langle s^0, s^1, \ldots \rangle$ is defined by $\pi^i = \langle s^i, s^{i+1}, \ldots \rangle$ and $\pi(i) = s^i$. The set of all paths is denoted as Π. Note that we do not require here that s^0 is an initial state of the Kripke structure.*

Temporal Logic

The introduced Kripke structures allow to consider linear time as well as branching time. In linear time each state has exactly one successor state, whereas in branching time several successors are possible. In model checking for both time models there is an according temporal logic, i.e. LTL for linear time and CTL for branching time, respectively. In the following the more expressive temporal logic CTL* is introduced, which contains both LTL and CTL.

Besides all the standard operators of propositional logic, CTL* contains two additional sets of operators. The linear time operators are for reasoning about a single path starting at the actual state: G expresses that a formula must hold for all successor states on the path, F indicates that a formula must be true for at least one of the successor states, X is used to reason about the immediate next state of the current state of the path and U states that a formula will be valid in all states until a second formula is true in some future states. The branching time operators are used to reason about sets of paths, i.e. the tree aspect in the Kripke structure is reflected. The operator A determines if a formula is true on all possible paths beginning at the current state. The dual operator of A is E which requires that a formula is true on one path only. A formal definition of the syntax of CTL* is given now.

DEFINITION 2.12 (SYNTAX OF CTL*) *Let V be the set of propositional variables (atomic formulas). Then, the two formula classes state formulas and path formulas are distinguished. They are defined as follows:*

State formulas:

- *Every $v \in V$ is a state formula.*
- *If p and q are state formulas, so are $\neg p$, $p \vee q$ and $p \wedge q$.*
- *If p is a path formula, then Ep and Ap are a state formulas.*

Path formulas:

- *If p is a state formula, then p is also a path formula.*
- *If p and q are path formulas, so are $\neg p$, $p \vee q$, $p \wedge q$, Xp, pUq, Fp and Gp.*

All state formulas are CTL formulas.*

Next, the semantics of CTL* is defined.

DEFINITION 2.13 (SEMANTICS OF CTL*) *Let $K = (S, S_0, R, V, L)$ be a Kripke structure, $s \in S$ a state and π a path in K. Furthermore, let q_1, q_2 be state formulas and p_1, p_2 be path formulas. For a state formula q, $K, s \models q$*

denotes that q holds at the state s of K. For a path formula p, $K, \pi \models p$ *denotes that p holds along the path* π *in* K*. If* K *is clear from the context it can be omitted. For the semantics of CTL* the relation* $\models$ *is inductively defined as follows:*

- $s \models v :\Leftrightarrow v \in L(s)$, *if* $v \in V$
- $s \models \neg q_1 :\Leftrightarrow$ *not* $s \models q_1$
- $s \models q_1 \vee q_2 :\Leftrightarrow s \models q_1$ *or* $s \models q_2$
- $s \models q_1 \wedge q_2 :\Leftrightarrow s \models q_1$ *and* $s \models q_2$
- $s \models Eq_1 :\Leftrightarrow$ *there exists a path* π *which starts at s, such that* $\pi \models q_1$
- $s \models Aq_1 :\Leftrightarrow$ *for all paths* π *starting at s, it holds that* $\pi \models q_1$
- $\pi \models q_1 :\Leftrightarrow s \models q_1$ *and s is the first state of* π
- $\pi \models \neg p_1 :\Leftrightarrow$ *not* $\pi \models p_1$
- $\pi \models p_1 \vee p_2 :\Leftrightarrow \pi \models p_1$ *or* $\pi \models p_2$
- $\pi \models p_1 \wedge p_2 :\Leftrightarrow \pi \models p_1$ *and* $\pi \models p_2$
- $\pi \models Xp_1 :\Leftrightarrow \pi^1 \models p_1$
- $\pi \models Gp_1 :\Leftrightarrow$ *for all* $k \geq 0$ *it holds that* $\pi^k \models p_1$
- $\pi \models Fp_1 :\Leftrightarrow$ *there exists a* $k \geq 0$ *such that* $\pi^k \models p_1$
- $\pi \models p_1 U p_2 :\Leftrightarrow$ *there exists a* $k \geq 0$ *such that* $\pi^k \models p_2$ *and for all* $0 \leq j < k$ *it holds that* $\pi^j \models p_1$

Based on the syntax and semantics introduced above we finally define when a CTL* formula holds for a Kripke structure.

DEFINITION 2.14 *A CTL* formula* φ holds *for a Kripke structure* $K = (S, S_0, R, V, L)$ *iff for all initial states* $s_0 \in S_0 : K, s_0 \models \varphi$.

As mentioned earlier CTL* covers the two most commonly used temporal logics:

- LTL is the subset obtained from CTL* by restricting the CTL* syntax to path formulas only, i.e. an LTL formula has the form Af, where f is a path formula which can only use atomic formulas as state formulas. By this, for LTL it is required that f has to be true on all possible paths beginning at an initial state. Consequently, the preceding A operator is implicitly assumed and not written for an LTL formula.

- CTL is obtained from CTL* by requiring that each temporal operator (X, G, F, U) must be directly preceded by a branching time operator A or E.

Frequently two types of formulas (or properties) are distinguished: *safety* properties and *liveness* properties. Informally speaking, safety properties express that "something bad does never happen", while liveness properties state that "something good eventually happens". An example for a safety property is mutual exclusion, as LTL property, e.g., $G\neg(ack_1 \wedge ack_2)$. An example for liveness is "any request will be granted eventually", as LTL property, e.g., $G(req \rightarrow Fack)$.

A *counter-example* to a safety property is a finite execution trace, while a counter-example to a liveness property is an infinite execution trace. For a finite-state system, such an infinite execution trace is a loop on states where the "good" behavior never happens.

Model Checking Algorithms

In this section the basics of model checking algorithms are reviewed. We start with a short presentation of *Symbolic Model Checking* (SMC). In SMC the states and the transitions are represented symbolically. Usually for this task BDDs are used, more precisely the transitions are represented by BDDs as well as the set of states by representing its characteristic function as a BDD [BCMD90, McM93, BCL+94]. To show whether a CTL formula holds for a given Kripke structure K, all states in K have to be determined in which the formula holds and it has to be checked if each initial state of K is contained in the computed state set. Technically, the state set computation for a CTL formula starts in the leaves of the formula and ends at its root. The cases at the leaves correspond to the evaluation of the labeling function for the variables at the leaves. In the implementation only the formulas $\neg\varphi, \varphi_1 \wedge \varphi_2, EX\varphi, E(\varphi_1 U \varphi_2)$ and $EG\varphi$ have to be handled with an according evaluation function since the remaining formulas can be reduced to these formulas. With the results from the leaves the propositional operators are calculated using the corresponding set operations. For the complex operators $E(\varphi_1 U \varphi_2)$ and $EG\varphi$ fix-point computations are necessary. The core steps for this task are the *image* and *pre-image* computations. These operations are used to determine the set of states that are reachable in one step from a given set of states and all predecessor states of the given set of states, respectively. Both operations are shown below:

$$\begin{aligned} Image(X) &= \{s' \in S \mid \exists s \in X \text{ with } R(s, s')\} \\ PreImage(X) &= \{s \in S \mid \exists s' \in X \text{ with } R(s, s')\}, \end{aligned}$$

where X is the given set of states. Both operations can be implemented efficiently using BDDs by carrying out conjunction and existential quantification on the respective BDD representations. However, for large designs these operations suffer from "memory explosion" problems.

Hence, methods for model checking have been developed to overcome this limitation. A very successful technique that is based on SAT is *Bounded Model Checking* (BMC). BMC has been introduced by Biere et al. in [BCCZ99] and gained popularity very fast. For industrial case studies see, e.g., [AKMM03, ADK$^+$05]. A survey in this context can be found in [PBG05] and for more details we refer the reader to [GG07]. For an LTL formula φ the basic idea of BMC is to search for counter-examples to φ in executions of the system whose length is bounded by k time steps. More formally, this can be described as:

$$BMC^k = I(S_0) \wedge \bigwedge_{i=0}^{k-1} T(s_i, s_{i+1}) \wedge \neg\varphi^k,$$

where $I(S_0)$ denotes the predicate for the initial states, T denotes the transition relation and $\neg\varphi^k$ constraints that the property φ is violated by an execution of length k.

Since LTL formulas have to hold for all paths, finding counter-examples means to check whether there exists an execution that contradicts the LTL formula. Such an execution is a witness for the negated LTL formula. For example, for a liveness property of the form AFp,[2] BMC tries to find a witness for $EG\neg p$, i.e. $\neg\varphi^k$ represents a loop within an execution of length at most k, such that p is violated on each state in the loop. BMC is also very often used to check safety properties. If the safety property has the form AGp where p is some propositional formula, then the considered witness is $EF\neg p$. Hence, in this case the resulting translation for $\neg\varphi^k$ is $\bigvee_{i=0}^{k} \neg p_i$, where p_i is the propositional formula at time point i. In general, a propositional formula has to be generated from the LTL formula φ for the given bound k. For the temporal operators of φ the well-known LTL expansion rules [MP91] are used, e.g., $Ff \equiv f \vee XFf$ and $Gf \equiv f \wedge XGf$. In addition, during the translation it is taken into account whether the witness requires an infinite behavior or not. For instance, a witness for EGp with a finite execution is only possible by including a back loop in the translation. For more details on the translation we refer the reader to [BCCZ99].

Finally, the overall problem formulation is transformed into CNF. Hence, if the resulting SAT instance is satisfiable a counter-example of length k has been found. Usually BMC is applied by iteratively increasing k until a counter-example for the property has been determined or the resources are

[2]For clarity the path quantifiers E and A are used to denote whether the formula has to hold for all paths or only for at least one path.

exceeded. For proving a property, k has to reach the sequential diameter which is not feasible for large designs. Therefore, approaches for BMC have been developed which can ensure completeness for safety properties (see, e.g., [SSS00, IPC03, WTSF04]), i.e. a property can be proven also. Note in [BAS02], Biere et al. have shown that liveness properties can be reduced to safety properties.

In this book a BMC variant is used that restricts the properties to bounded ones and thereby allows to prove a property. A detailed presentation of the approach including the syntax and semantics with respect to the used *Property Specification Language* (PSL) is given in Section 4.1.

2.4 SystemC

To handle the increasing complexity of circuits and systems the abstraction level for the design has been raised in the last decades: from transistors to gates, from gates to *Register Transfer Level* (RTL), and since a few years from RTL to C/C++ based models. This evolution has driven the development of the system description language SystemC [LTG97, GLMS02] which has been standardized by the IEEE [IEE05a]. SystemC has been introduced by the *Open SystemC Initiative* (OSCI) [OSC08], an independent, non-profit association composed of industrial and academic partners. In the following SystemC is briefly reviewed and a simple SystemC modeling example is provided.

2.4.1 Basics and Concepts

This section gives a brief introduction into the basics and concepts of the system description language SystemC. For a more detailed presentation we refer the reader to [GLMS02, MRR03, BD05].

The central idea of SystemC is to provide a language which spans from concept to implementation. Thus, in SystemC hardware and software systems can be modeled. Using traditional HDLs like VHDL or Verilog the simulation of hardware and software requires complex interfacing of hardware and software simulators. Thereby the simulation speed degrades and designers have to know both languages. Another advantage of SystemC is that the initial high level description of a system can be refined progressively within one language trough the different levels of abstraction, until finally a synthesizable description is reached. The underlying SystemC refinement methodology eliminates the manual translation step into a traditional HDL.

For modeling hardware the following aspects are fundamental:

- Concurrency, since hardware is inherently concurrent.
- A notion of time is required.
- HW datatypes to support for example tristates.

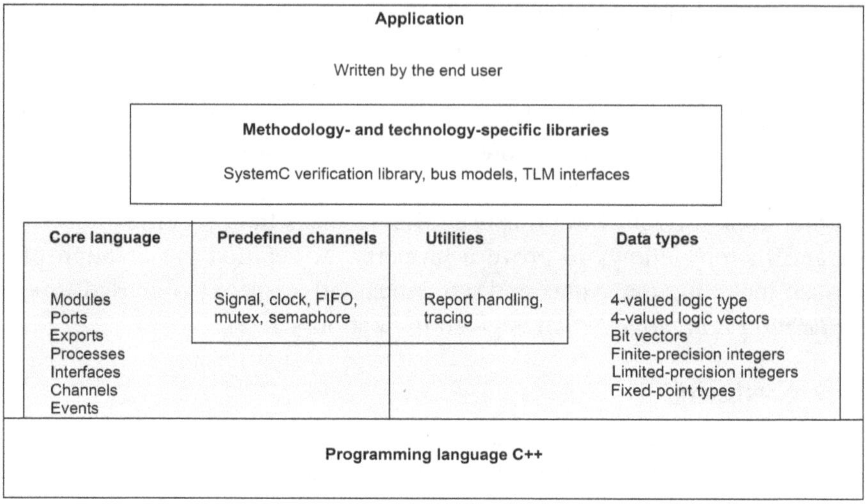

Figure 2.6. Architecture of SystemC [IEE05a]

All these aspects have been incorporated during the development of SystemC. Essentially, SystemC is a C++ class library. Thus, SystemC inherits all features of C++, e.g., the concept of object orientation. By this, the issues for describing software are taken into account. For the description of hardware in addition to the above mentioned aspects the SystemC class library adds a simulation semantic to C++ which is similar to the one of event-based HDL simulators. This includes the concept of delta cycles as well as sensitivity lists.

The architecture of SystemC is depicted in Figure 2.6. On top of C++ the SystemC class library is shown as shaded blocks. The class library consists of four parts: the core language, the SystemC data types, the predefined channels and the utilities. The layer above SystemC represents standard or proprietary C++ libraries that provide specific design or verification methodologies. One example is the add-on library to support *Transaction Level Modeling* (TLM) [CG03]. TLM allows to describe the communication in a system in terms of abstract operations (transactions). Another example is the *SystemC Verification* (SCV) library [Sys03] which is considered in more detail in the next chapter. In the following we summarize the major parts of the SystemC layer with the focus on modeling a system:

- Modules are the basic building blocks for partitioning a design. A module can contain processes, ports, channels and other modules. Thus, a hierarchical design description becomes possible.

- Communication is realized with the concept of interfaces, ports and channels. An interface defines a set of methods to access channels. Through

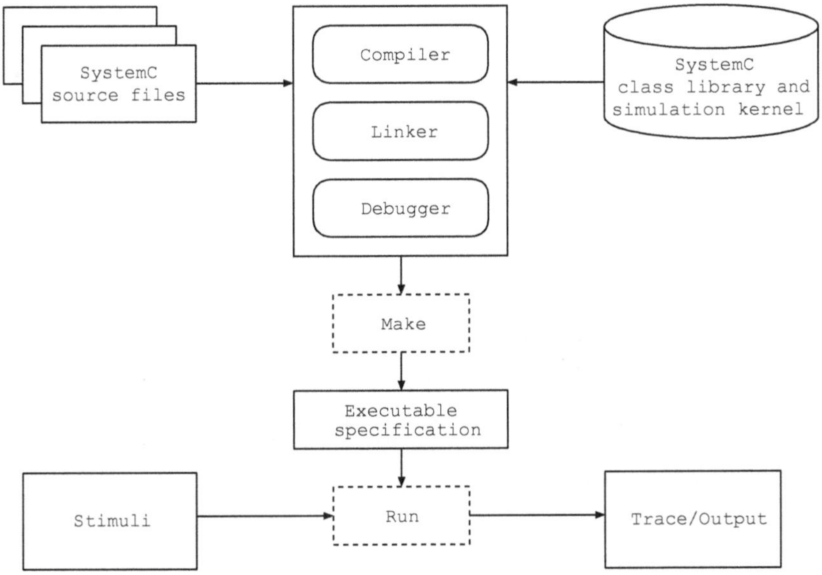

Figure 2.7. SystemC tool flow

ports a module can send or receive data and access channel interfaces. A channel serves as a container for communication functionality, e.g., to hide communication protocols from modules.

- Processes are used to describe the functionality of the system and allow expressing concurrency in the system. They are declared as special functions of modules and can be sensitive to events, e.g., an event on an input signal.

- Hardware specific objects are supplied, like, e.g., signals, which represent physical wires, clocks, and a set of data-types useful for hardware modeling.

Besides this, SystemC provides a simulation kernel. Note that a SystemC description can be compiled with a standard C++ compiler to produce the *executable specification* of the system. By starting the executable the system is simulated. The resulting tool flow for simulation is illustrated in Figure 2.7. In the next section a simple SystemC example is given. More complex ones are presented in the respective chapters.

2.4.2 SystemC Design Example

In the following a simple SystemC description is presented to exemplify the above mentioned modeling features.

```
1  typedef sc_uint<3> T;
2
3  SC_MODULE(counter) {
4    // ports and signals
5    sc_in_clk clock;
6    sc_in<bool> reset;
7    sc_out< T > out;
8    T count_val;
9
10   // processes
11   void do_count() {
12     if (reset.read()) {
13       count_val = 0;
14     } else {
15       count_val = count_val + 1;
16     }
17     out = count_val;
18   }
19
20   // constructor
21   SC_CTOR(counter) {
22     SC_METHOD(do_count);
23     sensitive << clock.pos();
24   }
25 };
```

Figure 2.8. SystemC counter

EXAMPLE 2.15 *In Figure 2.8 a SystemC implementation of a counter is shown. The bit size of the counter is determined by a* typedef. *Here, the type* T *is defined as an unsigned integer with 3 bits (line 1). The SystemC module* counter *has two inputs,* reset *and* clock. *In addition, the counter has one output* out. *The current value of the counter is stored in* `count_val`. *In the process* do_count *the functionality of the counter is described. Obviously the counter counts modulo 8 based on the given typedef. Finally, in the constructor of the module the method* do_count *is declared as an* SC_METHOD *which is sensitive to the positive clock.*

To simulate the SystemC counter a testbench that provides the input stimuli and a simulation environment are required. The simulation environment is provided by specifying the top level function `sc_main()`. Here, the modules of the design are instantiated, connected and the simulation is started.

EXAMPLE 2.16 *The respective* sc_main *for the counter is shown in Figure 2.9. After declaration of signals a module for stimuli generation (line 10) and*

```
1  #include<systemc.h>
2  #include "counter.h"
3  #include "stimGen.h"
4
5  int sc_main(int argc, char *argv[]) {
6    sc_clock clk("Clock", 1, 0.5, 0.0);
7    sc_signal < bool > reset;
8    sc_signal < T > out;
9
10   stimGen s("Stimuli␣Generator");
11   s.start = 5;
12   s.reset(reset);
13   s.clock(clk);
14
15   counter c("Counter");
16   c.clock(clk);
17   c.reset(reset);
18   c.out(out);
19
20   sc_trace_file *tf = sc_create_vcd_trace_file("counter");
21   sc_trace(tf, reset, "reset");
22   sc_trace(tf, out, "out");
23
24   // start simulation
25   sc_start(clk, 100);
26
27   sc_close_vcd_trace_file(tf);
28   return 0;
29 }
```

Figure 2.9. Top function `sc_main`

the counter (line 15) are instantiated. The implementation of the module for stimuli generation is not given. The connections (or port bindings) are accomplished after module instantiation. For example, in line 17 the input reset of the counter is bound to the above declared reset signal. Then, a trace file is created and the reset signal and the out signal are recorded. Finally, the counter is simulated for 100 clock cycles (line 25). The resulting waveform is shown in Figure 2.10. The reset is released after the fifth clock cycle (the waveform viewer dinotrace [Sny08] starts to count from zero).

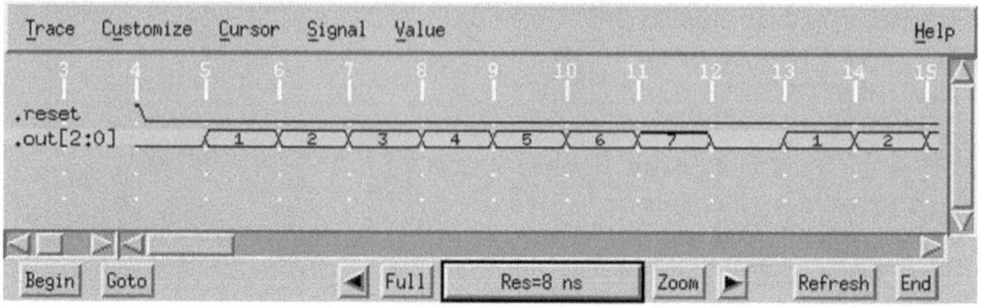

Figure 2.10. Waveform for simulated SystemC counter

Chapter 3

SYSTEM-LEVEL VERIFICATION

The parts of the proposed design and verification flow covered in this chapter are shown in Figure 3.1. As already mentioned, for modeling a system in this book the system description language SystemC is used. Thus, from the textual specification the initial system-level model is directly described in SystemC. Following the design methodology of SystemC the system-level model is very abstract and can be simulated at this high level of abstraction already by compiling the model into the executable specification. This allows for efficient design space exploration. After analyzing the results of a certain design direction the designer can go back and revise design decisions (for simplicity this loop is not shown in the figure). Since not all details have been modeled already, this can be accomplished with moderate costs. Also part of the design space exploration phase is to check hardware/software trade-offs. Hence, hardware/software partitioning is performed to meet the requirements of the specification. During the development of the system-level model verification is started.

In the middle of Figure 3.1 the dedicated verification techniques and the respective quality check that are proposed in this chapter for the system level are depicted. In the first part of this chapter constraint-based simulation is considered which overcomes the limitations of "pure" simulation. Constraint-based simulation is based on stimulus generation by constraint solving. The resulting stimuli will in particular cover corner case test scenarios which are usually hard to identify manually by the verification engineer. For SystemC the *SystemC Verification* (SCV) library [Sys03] offers constraint-based simulation. The underlying constraint-solver of the SCV library is based on formal methods. More precisely, *Binary Decision Diagrams* (BDDs) are used to represent the constraints. In Section 3.1 the scenario of constraint-based simulation in

DOI 10.1007/978-90-481-3631-5_3,

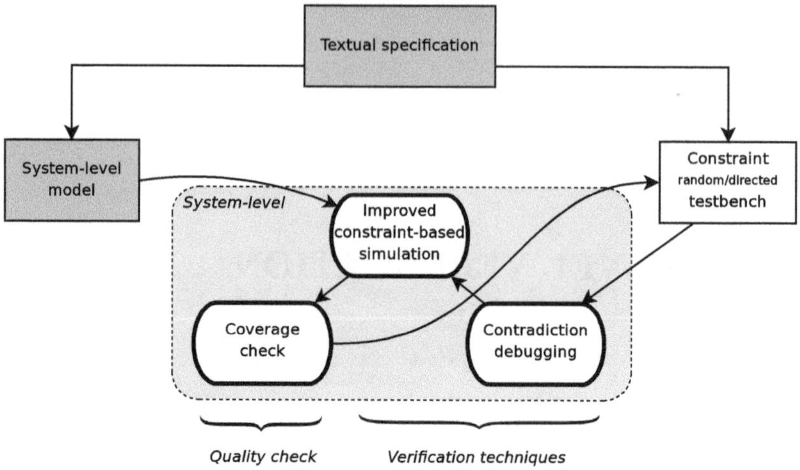

Figure 3.1. System-level parts of enhanced SystemC design and verification flow

general and thereafter in the context of SystemC using the SCV library is considered. Section 3.2 presents two improvements: First, new operators for the specification of SCV constraints are provided. Second, the uniform distribution among all solutions of a constraint is guaranteed for maximizing the chance of entering unexplored regions of the design state space. Both improvements have been published in [GED07]. In Figure 3.1 they are summarized in the verification task *improved constraint-based simulation*.

While specifying complex non-trivial constraints (and for example focussing on special scenarios), the verification engineer is faced with the problem of over-constraining, i.e. the overall constraint defined for a certain test scenario has no solution. In this case the root cause of the contradiction has to be identified and resolved. Given the complexity of constraints used to describe test scenarios, this can be a very time-consuming process. Thus, in Section 3.3 a fully automated approach for contradiction analysis is introduced. The method determines all "non-relevant" constraints and computes all reasons that lead to the over-constraining (see task *contradiction debugging* in the figure). The approach has been published in [GWSD08].

Finally, to ensure the resulting verification quality we investigate the question how thorough the design was tested. Hence, in Section 3.4 an approach to measure the quality of the testbench is presented. Dedicated code coverage techniques that have been developed for SystemC models are used for the analysis. As a result a coverage report is generated that shows all statements in the model that have not been executed during simulation (see task *coverage check* in Figure 3.1). The method has been published in [GPKD08].

3.1 Constraint-Based Simulation

First, in this section the general scenario of constraint-based simulation is described. Then, the SystemC verification library is introduced with the particular focus on constraint-based simulation for SystemC designs.

3.1.1 Scenario

Constraint-based simulation is used to improve pure simulation. Constraint-based simulation can be used in early design phases as well as in subsequent phases where the design is stepwise refined down to synthesizable descriptions. In the following the difference of constraint-based simulation in comparison to pure simulation is explained.

In *directed simulation*[1] explicitly specified stimulus patterns, e.g. written by verification engineers, are applied to the design. Each of those patterns stimulates a very specific design functionality, called a verification scenario, and the response of the design is compared thereafter with the expected result. Due to project time constraints, it is inherent for directed simulation that only a limited number of such scenarios can be verified.

With *random simulation* these limitations are compensated. Random stimuli are generated as inputs for the design. For example, to verify the communication over a bus, random addresses and random data are computed.

A substantial time reduction for the creation of simulation scenarios is achieved by *constraint-based simulation* (see e.g. [YSP+99, Ber06, YPA06]). Here, the stimuli are generated directly from specified constraints by means of a constraint solver, i.e. stimulus patterns are selected by the solver which satisfy the constraints. The resulting stimuli will also cover test scenarios for corner cases that may be difficult to generate manually. As a consequence, design bugs will be found that might otherwise remain undetected, and the quality of design verification increases substantially. In the next section the SystemC verification library is introduced with a focus on constraint-based simulation.

3.1.2 Using the SCV Library

The first version of the *SystemC Verification* (SCV) library was introduced in 2002 as an open source class library [Sys03, RS03, IS03]. The SCV library layers on top of SystemC (see also Section 2.4) and adds tightly integrated verification capabilities to SystemC. In the following the main features of the SCV library are summarized:

[1]Directed simulation [BMA05] is typically seen as pure simulation, since it is the simplest form of simulation.

- Data introspection for SystemC and C++ data types
  - Manipulation of arbitrary data types
  - Manipulation of data objects without compile time information
- Transaction API
  - Transaction monitoring and recording
  - Basis for debugging, visualization and coverage by providing transaction data
- Constraint-based stimulus generation for SystemC and C++
  - High quality pseudo random generator
  - Integrated constraint solver based on BDDs

Since we present in the next sections several improvements for constraint-based simulation with the SCV library, we give in the following some examples how constraints are specified using this library. In addition, the representation of the constraints using BDDs is discussed.

Before the SystemC specific concepts for constraint specification are described, the type and structure of constraints is formalized in the following definitions.

DEFINITION 3.1 *A* constraint *is a Boolean function over variables from the set of variables V. For the specification of a constraint the typical HDL operators such as, e.g. logic AND, logic OR, arithmetic operators and relational operators can be used.*

Usually a constraint consists of a conjunction of other constraints. We formalize the resulting *overall constraint* in the following definition.

DEFINITION 3.2 *An* overall constraint *is defined as*

$$C = \bigwedge_{i=0}^{n-1} C_i$$

where C_i are constraints according to Definition 3.1.

Within the SCV library, constraints are modeled in terms of C++ classes. The constraints can be hierarchically layered using C++ class inheritance. In detail a constraint is derived from the `scv_constraint_base` class. The data to be randomized is specified as `scv_smart_ptr` variables. The conjunction as shown in Definition 3.2 is built in SystemC by the explicit use of several SCV_CONSTRAINT() macros or by applying inheritance, i.e. parts of the constraints are defined in a base class and inherited by the actual constraint. Note

```
struct my_constraint : public scv_constraint_base {
  scv_smart_ptr<sc_uint<64> > a, b, addr;

  SCV_CONSTRAINT_CTOR( my_constraint ) {
    SCV_CONSTRAINT( a() > 100 );
    SCV_CONSTRAINT( b() == 0 );
    SCV_CONSTRAINT( addr() >= 0 && addr() <= 0x400 );
  }
};
```

Figure 3.2. Example constraint

that this is not specific to constraint-based simulation using the SCV library. In fact, the same principles are found, for example, in the random constraints of SystemVerilog [IEE05b].

We illustrate the specification of SCV constraints in the following.

EXAMPLE 3.3 *An example of an SCV constraint is shown in Figure 3.2. The name of the constraint is* my_constraint. *Here, the three 64-bit unsigned integer variables a, b and addr are randomized. The conditions on the variables a, b and addr are defined by expressions in the respective* SCV_CONSTRAINT() *macro. For example, the value of the variable a is restricted to be always greater than* 100. *All constraint expressions specified by the respective* SCV_CONSTRAINT() *macro have to hold in conjunction, i.e. the underlying constraint solver computes a logical AND of all the constraint expressions.*

The above mentioned principle of hierarchical constraints is shown in the next example.

EXAMPLE 3.4 *Using the C++ concept of inheritance the hierarchical constraint in Figure 3.3 is formulated. From the logical point of view the overall constraint consists of the conjunction of the base constraint and the current constraint expressions given in the* hierarchical_constraint. *The constraint expressions from the base class are "imported" to the derived class with the macro* SCV_CONSTRAINT_BASE(). *In summary, the* hierarchical_constraint *adds another restriction to* my_constraint, *i.e. the value of a has to be smaller than* 32768.

It is also possible to merge more than one constraint by inheriting from the respective constraint classes. In addition, the derived constraint can also define new constraint variables.

In the SCV library a constraint is represented by the corresponding characteristic function, i.e. the function is true for all solutions of the constraint. This

```
struct hierarchical_constraint : public my_constraint {

  SCV_CONSTRAINT_CTOR( hierarchical_constraint ) {
    SCV_CONSTRAINT_BASE( my_constraint );
    SCV_CONSTRAINT( a() < 32768 );
  }
};
```

Figure 3.3. Hierarchical constraint

```
1 struct simple : public scv_constraint_base {
2   scv_smart_ptr<sc_uint<4> > a;
3
4   SCV_CONSTRAINT_CTOR( simple ) {
5     SCV_CONSTRAINT( a() > 0 );
6   }
7 };
```

Figure 3.4. Simple constraint

characteristic function of a constraint is represented as a BDD.[2] The following example illustrates a constraint and the corresponding BDD representation.

EXAMPLE 3.5 *In Figure 3.4 the specification of the SCV constraint* simple *is shown. This constraint uses only a single variable of bit width 4 (line 2) and a simple constraint expression (line 5) so that the corresponding BDD has a small size and can be understood easily.*[3] *The characteristic function of this constraint represented as a BDD is depicted in Figure 3.5. The constraint variable a is equivalent to $a_0a_1a_2a_3$ where a_0 is the LSB of a. In the figure the 0-edge is shown as a dashed line and a dot on an edge represents a complement edge (i.e. the function below is inverted). As can be seen the BDD evaluates to* 0 *for the assignment $a_0a_1a_2a_3 = 0000_2 = 0_{10}$ only, all other assignments lead to* 1. *Obviously this BDD represents the* simple *constraint.*

The presented information in this section mainly focused on the specification of constraints using the SCV library. In the next section, the available operators in SCV constraints are enriched. Furthermore, constraint solving with respect to a uniform distribution among all solutions of a constraint is considered in detail.

[2]The BDD package CUDD [Som98] is used in the SCV library.

[3]If more variables are used especially the interleaved variable ordering makes it hard to see the relation of the constraint and the BDD representation of its characteristic function.

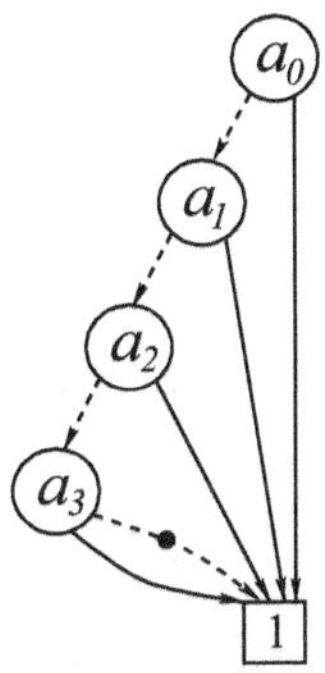

Figure 3.5. BDD of simple constraint

3.2 Improvements for Constraint-Based Simulation

The improvements for constraint-based simulation using the SCV library are motivated by observations in an industrial setting. At the AMD Dresden Design Center (DDC) a verification environment that integrates the constraint-based simulation of the SCV library is used. There it has been observed that the SCV library has two major disadvantages which restrict the practical use. On the one hand in the constraints no bit operators are supported. These are important if parts of the system have been refined already to the lower levels. On the other hand the constraint solver does not fulfill the important requirement that the constraint solutions are distributed uniformly if certain constraint variables are set to a fixed value, i.e. the randomization for these variables is disabled. We analyze these two problems and describe improvements that overcome these limitations.

The section is structured as follows. In Section 3.2.1 new bit operators for SCV constraints are introduced that significantly help during constraint specification. Section 3.2.2 explains how a uniform distribution across all constraint solutions can be guaranteed.

3.2.1 Bit Operators

In Section 3.1.2 the specification of constraints using the SCV library including several examples has been presented. In the following the composition of constraint expressions is discussed in more detail. Constraint expressions over variables to be randomized can only use the following operators [Sys03]:

- Arithmetic operators: `+`, `-`, `*`
- Relational operators: `==`, `!=`, `>`, `>=`, `<`, `<=`
- Logical operators: `!`, `&&`, `||`

```
1 struct constraint : public scv_constraint_base {
2   scv_smart_ptr<sc_uint<64> > address;
3   scv_smart_ptr<sc_uint<32> > mode;
4
5   SCV_CONSTRAINT_CTOR(constraint) {
6     SCV_CONSTRAINT( address().range(0,1) == 0 );
7     SCV_CONSTRAINT( mode()[10] == 1 );
8   }
9 };
```

Figure 3.6. Example constraint with bit operators

As can be seen there is no support for bit operators in the SCV constraint solver. However, bit operators are very important for the verification engineer during the specification of constraints. Bit operators allow for simpler and more compact formulations of complex constraints. In detail the following bit operators have been added to the SCV library:

1. Bitwise AND: `a() & b()`
2. Bitwise OR: `a() | b()`
3. Bitwise NOT: `~a()`
4. Bit-select: `a()[i]` for constant i
5. Slice-select: `a().range(x,y)` for constant x and y

A concrete example shows the usefulness of the new operators.

EXAMPLE 3.6 *In Figure 3.6 an example constraint that uses bit operators is depicted. For the 64-bit variable* address *the constraint enforces that the two lower bits are* 0 *(line 6). Such a specification is typically used to express valid word-wise addressing in a constraint. Without the new* range(x,y) *operator this can only be formulated by explicitly enumerating all allowed values, i.e.* address()==0 || address()==4 || address()==8 ... *Such an enumeration is obviously not desirable for large address variables as it is the case in the example. A similar argumentation holds for setting a single bit in the* mode *variable.*

A more complex industrial constraint example is presented in Section 3.3 where PCI express constraints are described. In addition, this example makes use of the "bitwise AND" operator.

For the integration of the new operators in the SCV library, first in the class `scv_expression` the according C++ operators have been overloaded and new member functions have been added. The class `scv_expression` is used for the internal representation of the constraint expressions in form of an expression tree. In such a tree leaf nodes are variables or constants and non-terminal nodes are marked with operators. The class `_scv_expr` is used to store the BDD representation of an `scv_expression`. For the construction of the BDD in this class each bit operator has to be mapped to the corresponding BDD synthesis operation. For example, in case of a "bitwise AND" the resulting bit vector is computed by the BDD-AND operation for each bit of the two input vectors. Of course there are several special cases like different length of vectors, different data types etc. that have to be taken into account.

In summary, the new operators allow a concise and compact way of constraint specification.

3.2.2 Uniform Distribution

The uniform distribution of the solutions of constraints is a very important aspect for the quality of a constraint solver. However, this is not supported by the SCV library in all circumstances. The problem occurs in scenarios of high practical relevance. If variables are fixed to a certain value for constraint solving, i.e. these variables are disabled for randomization, then the solutions computed by the constraint solver are not uniformly distributed across the set of all possible solutions. Later this phenomenon will be illustrated by a simple example. Before describing the solution to this problem the constraint solving process of the SCV library is explained in more detail.

The constraint solver works on individual bits when solving constraints. As explained in Section 3.1.2 for a constraint, a BDD representation is computed. The constraint solver generates a solution of a constraint by using the BDD that represents the constraint. For this purpose the algorithm starts at the root node and traverses the BDD down to the 1-terminal. A path starting from the root and ending at the 1-terminal determines the values of the variables along the path. These values correspond to a solution of the constraints since the BDD is the characteristic function of the constraint. One could assume that choosing the 0-assignment or 1-assignment for a Boolean variable with a probability of 50% guarantees a uniform distribution. However, the following observation shows that this is not true. As explained above all constraint solutions are paths to the 1-terminal starting from the root node. But during the BDD traversal some sub-BDDs can have more paths to the 1-terminal than other sub-BDDs. Thus, if a sub-BDD with fewer paths is selected this leads to an overweighting of the fewer represented solutions. This is illustrated by the following example.

EXAMPLE 3.7 *In Figure 3.7 the BDD for the function* $f = \overline{x}_1 x_2 + x_1 x_2 x_3$ *is shown, where dashed lines (solid lines) are used for the 0-assignment (1-assignment) and a dot on an edge represents a complement edge (i.e. the*

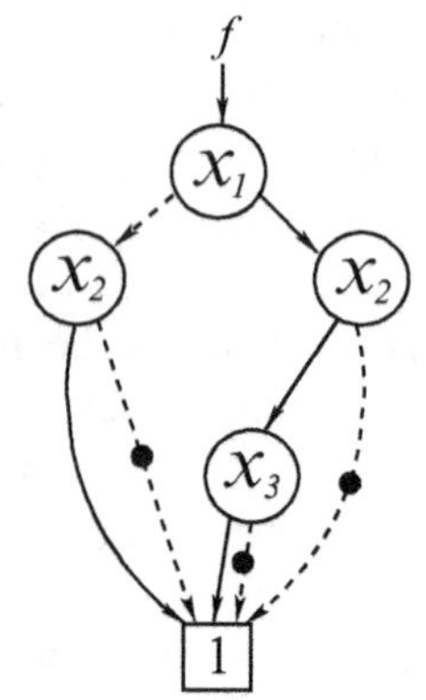

Figure 3.7. BDD for $f = \overline{x}_1 x_2 + x_1 x_2 x_3$

Table 3.1. Probabilities for solutions

x_1	x_2	x_3	probability
0	1	0	25%
0	1	1	25%
1	1	1	50%

function below is inverted). If during the BDD traversal of the function the 1-edge of the root node is chosen there is exactly one path to the 1-terminal. If instead the 0-edge is chosen, the reached sub-BDD has two *paths to the 1-terminal: In the non-reduced BDD there is a node marked with x_3 that is reached by assigning $x_1 = 0$ and $x_2 = 1$; the 1-edge of this node as well as the uncomplemented 0-edge point to the 1-terminal. In total, the probabilities following the intuitive traversal algorithm are shown in Table 3.1.*

This example demonstrates that the probability for choosing the 1-edge of the root node should be corrected to 33% instead of 50%. By this, a uniform distribution across all solutions is achieved.

In the SCV constraint solver a special weighting algorithm is implemented to guarantee the uniform distribution of all solutions. In a pre-processing step the BDD of all initial constraints is traversed and the correct probabilities are computed for each node. The basic idea of the recursive weighting algorithm is to compute weights of the else- and then-child of a node while taking into account whether nodes have been removed due to BDD reduction rules. Based on the weights a probability is assigned to each BDD node. Then, for the generation of values – one solution is picked uniformly distributed across all solutions of the constraint – the computed probabilities of the pre-processing step are used during the BDD traversal.

The SCV constraint solver calls the weighting algorithm only once at the beginning for the initial BDD that represents the constraints (below the reasons including the required modifications of the SCV data structures are given). Thus, the SCV constraint solver is not able to handle simplifications like e.g. fixing variables to a certain value. In this case the BDD that represents the constraints is modified due to the simplification but the probabilities are not updated. This causes a non-uniform distribution of the constraint solution. We provide an example for this observation. For simplicity we only use one single constraint in the following example. Note that the presented technique is not restricted to this case. An arbitrary number of constraints as well as derived constraints are fully supported. After the example some technical details are given for solving the problem.

EXAMPLE 3.8 *Consider the constraint in Figure 3.8. This constraint specifies that* $a+b = c$ *and* c *is fixed to* 99 *(see lines 7 and 8). The distribution shown in Figure 3.9 was the result from running this constraint 100,000 times in the SCV constraint solver. As can be seen there is a strong bias of the solutions in the middle part.*

One has to overcome several difficulties while correcting this behavior. This is due to the design of the SCV library which divides the functionality of handling the BDD-based representation of constraints roughly into two parts, i.e. one global constraint manager object and the respective constraint objects (one for each constraint specified). On the one hand, it may seem natural to direct all BDD-related tasks via one dedicated constraint manager object (which is encapsulated in a class called `_scv_constraint_manager` and creates a manager object for CUDD at start). On the other hand, a closer inspection unveils serious flaws in the centralized design:

The SCV constraint manager maintains the number of BDD variables necessary for representing the least recently used constraint object. However,

```
1  struct triangle_c : public scv_constraint_base {
2    scv_smart_ptr<sc_uint<7> > a,b;
3    scv_smart_ptr<sc_uint<8> > c;
4
5    SCV_CONSTRAINT_CTOR(triangle_c) {
6      SCV_CONSTRAINT( a() + b() == c() );
7      c->disable_randomization();
8      *c = 99;
9    }
10 };
```

Figure 3.8. Triangle constraint

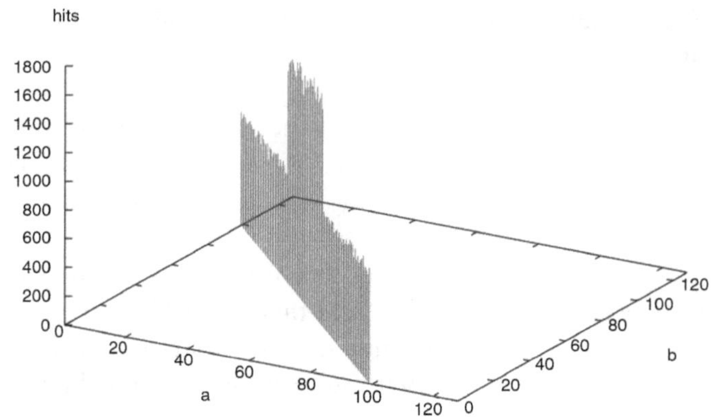

Figure 3.9. Distribution for $a + b = 99$ with original SCV

this number is reset, i.e. "forgotten", as soon as a new constraint object is created. Hence, if a previous constraint is simplified, this number is not available anymore. This is a problem since the weighting algorithm crucially depends on this number. There are similar problems with the information contained in two hash tables stored at the constraint manager object: the tables `nodeHashP` and `nodeWeightHash` hold the probability information for all nodes of the BDD representing a constraint and the according weighting information, respectively. At the time of simplification of a previous constraint, the data stored in the table needs to be cleared which is not done by the SCV system.

We did a complete redesign of the constraint management classes that also solved these problems. Furthermore, we put more intelligence into the constraint objects. For example, now every constraint object is capable to return a pointer to its BDD representation via a method `getBddNodeP`. Backwards compatibility has been preserved and full functionality/structure of the SCV interfaces, e.g. the possibility of overloading C++ virtual member functions like `scv_constraint_base::next` which triggers the next random assignment of the constraint variables. In order to achieve this, it was necessary to give the simplified BDD to the constraint object in several methods of the class `_scv_constraint_manager`, e.g. `assignRandomValue`, as well as in several internal utility routines called from other code within the SystemC verification standard, e.g. in `_scv_set_value`.

The result of our redesign now is a tight integration of the weighting algorithm and the BDD synthesis operations. After structural modifications in the interfaces and correct initialization of internal data structures now the weighting algorithm is called after a simplification. Thus, the weights and probabilities are recomputed and a uniform distribution is achieved. In Figure 3.10 the

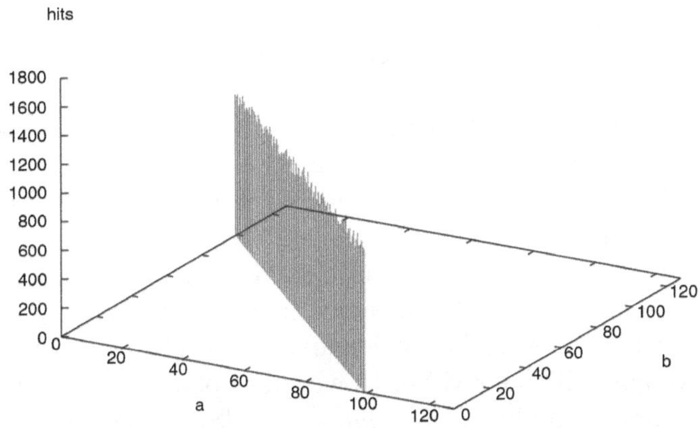

Figure 3.10. Distribution for $a + b = 99$ with improved SCV

result for the constraint from Example 3.8 is shown again for 100,000 times calling the SCV constraint solver. As can be seen a uniform distribution among all solution was established and hence the chance for entering unexplored regions of the design state space is maximized.

3.3 Contradiction Analysis for Constraint-Based Simulation

For constraint-based random simulation several approaches have been proposed (see e.g. [YSP+99, YAPA04, DKBE02, Iye03, KK07]). However, besides concise specifications of constraints and a uniform distribution across all constraint solutions as discussed in the previous section, a crucial problem arises: If complex constraints are specified for a test scenario, the verification engineer can be faced with *over-constraining*, i.e. the constraint solver is not able to find a valid solution for the given set of constraints. Whenever such a contradiction occurs in a constraint-based simulation run, this run has to be terminated as no valid stimulus patterns can be applied. Note that over-constraining may not necessarily happen at the very beginning of the simulation run, as modern testbench languages allow the addition of constraints dynamically during simulation. In any case of over-constraining the verification engineer has to identify the root cause of the contradiction. As this is usually done manually by either code inspection or trial-and-error debug, it is a tedious and time-consuming process.

To the best of our knowledge in this book we propose the first non-trivial algorithm for contradiction analysis in constraint-based simulation. In the area of constraint satisfaction problems methods for diagnosing over-constrained problems have been studied (see e.g. [BDTW93, PRB01]). These methods aim to find a solution for the over-constrained problem by relaxing constraints according to a given weight for each constraint. In the considered problem no weights are available. Also, the approaches do not determine all minimal reasons that cause the overall contradiction. In contrast, Yuan et al. proposed an approach to locate the source of a conflict using a kind of exhaustive enumeration [YPA06]. But since a very large run-time of this method is supposed – neither an implementation nor experiments are provided – they recommend to build an approximation. In the domain of *Boolean Satisfiability* (SAT) a somewhat similar problem can be found: computing an UNSAT core of an unsatisfiable formula, i.e. to identify an unsatisfiable sub-formula of the overall formula [GN03, ZM03]. However, to obtain a *minimal* reason the much more complex problem of a *minimal* UNSAT core has to be considered [OMA+04, Hua05, MLA+05]. Furthermore, all minimal UNSAT cores are required to determine all minimal contradictions. In general this is very time consuming (see e.g. [LS05]).

In this section we propose a fully automatic technique for analyzing contradictions in constraint-based random simulation. The basic idea is as follows: The overall constraint is reformulated such that (contradicting) constraints can be disabled by introducing new free variables. Next, an abstraction is computed that forms the basis for the following steps. First, the self-contradicting constraints are identified. Then, all "non-relevant" constraints are determined. Finally, for the remaining constraints – typically only a very small set – a detailed analysis is performed. In total our approach identifies *all* reasons of the over-constraining, i.e. all minimal constraint combinations that lead to a contradiction of the overall constraint. The proposed technique has been evaluated in a verification environment at AMD Dresden Design Center (DDC). As shown by the experiments the debugging time is reduced significantly. The verification engineer can completely understand what causes the over-constraining and thus can resolve the contradictions in one single step.

This section is structured as follows. In Section 3.3.1 the considered problem is formalized and the concepts of the contradiction analysis approach are given. The implementation of the approach is described in detail in Section 3.3.2. Section 3.3.3 provides experimental results. First, some results obtained for different types of contradictions are discussed. Then, the application of the analysis technique for a real-life industrial example in a verification environment at AMD DDC is presented.

3.3.1 Contradiction Analysis Approach

In this section first the considered problem is formalized. Then, the concepts for the contradiction analysis approach are presented.

Problem Formulation

During the specification of complex non-trivial constraints, the problem of over-constraining arises:

DEFINITION 3.9 *An overall constraint* $C = \bigwedge_{i=0}^{n-1} C_i$ *is* over-constrained *or* contradictory *iff C is not satisfiable, i.e. C always evaluates to* 0.

Typically, if C is over-constrained the verification engineer has to manually identify the reason for the over-constraining. This process can be very time-consuming because several cases are possible. For example, one of the constraints C_i may have no solution. Another reason for a contradiction may be that the conjunction of some of the constraints C_i leads to 0. In the following the term *reason* is defined as used in the remaining part of this section.

DEFINITION 3.10 *A* reason *for a contradictory overall constraint* C *is the set* $R = \{C_{i_1}, C_{i_2}, \ldots, C_{i_k}\} \subseteq \{C_0, C_1, \ldots, C_{n-1}\}$ *with the two properties:*

1. *The constraints in R form a contradiction, i.e. the conjunction* $C_{i_1} \wedge C_{i_2} \wedge \ldots \wedge C_{i_k}$ *always evaluates to* 0. *Therefore the overall constraint C is over-constrained.*
2. *Removing an arbitrary constraint from R resolves the contradiction, i.e. minimality of R is required.*

Often the root cause of over-constraining can result from more than one contradiction, i.e. there is more than one reason. If in this case only one reason is identified by the verification engineer, the constraint solver has to solve the fixed constraint again, but still there is no solution.

Based on these observations, the following problem is considered here:

How can we efficiently compute all minimal reasons for an over-constraining and thereby support the verification engineer in constraint debugging?

Analyzing the contradictions in the overall constraint C and presenting all reasons is facilitated by our approach. In particular excluding all constraints which are not part of a contradiction reduces the debugging time significantly.

Concepts for Contradiction Analysis

The general idea of the contradiction analysis approach is as follows: The overall constraint C is reformulated such that the conflicting constraints can be disabled by the constraint solver and C becomes satisfiable. By analyzing the logical dependencies of the disabled constraints, we can identify *all* reasons for the over-constraining.

DEFINITION 3.11 *Let C be over-constrained. Then the* reformulated constraint C' *is built by introducing a new free variable s_i for each constraint C_i and by substituting each constraint C_i with an implication from s_i to C_i. That is,*

$$C' = \bigwedge_{i=0}^{n-1} (s_i \rightarrow C_i).$$

For the reformulated constraint C' the following holds:

1. If s_i is set to 1, then the constraint C_i is enabled.
2. If s_i is set to 0, then the constraint C_i is disabled because C_i may evaluate to 0 or 1.

Note that the usage of an implication is crucial. If an equivalence is used instead of an implication, $s_i = 0$ would imply the negation of C_i.

EXAMPLE 3.12 *Figure 3.11 shows a constraint C which is over-constrained (for clarity the SCV specific declaration parts are not shown, in fact each constraint expression is marked with C_i). Reformulating C to C' avoids the over-constraining because a constraint C_i may be disabled by assigning s_i to 0. Table 3.2 gives all assignments to s_i such that the reformulated overall constraint C' evaluates to 1.*[4] *That is, the table shows which constraints have to be disabled to get a valid solution. For example, disabling C_0, C_2, C_3 and C_5 avoids the contradiction.*

$C_0 \Leftrightarrow$ b()<3 && b() == 7

$C_1 \Leftrightarrow$ a() + b() == c()

$C_2 \Leftrightarrow$ a() < 6

$C_3 \Leftrightarrow$ a() == 5

$C_4 \Leftrightarrow$ a() == 10

$C_5 \Leftrightarrow$ d() == 8

$C_6 \Leftrightarrow$ d() > 10

Note that all variables a(), b(), c(), d() are positive integers.

Figure 3.11. Contradictory constraint

[4]Here '$-$' denotes a don't care, i.e. the value of s_i can be either 0 or 1. The table is derived from a symbolic BDD representation of all solutions for the s_i variables after abstraction of all other variables.

Table 3.2. Table of contradictory constraint

s_0	s_1	s_2	s_3	s_4	s_5	s_6
0	–	0	0	–	0	–
0	–	0	0	–	1	0
0	–	0	1	0	0	–
0	–	0	1	0	1	0
0	–	1	–	0	0	–
0	–	1	–	0	1	0

Based on the reformulation the verification engineer is able to avoid the over-constraining. However, in order to understand what causes the over-constraining, i.e. to identify the reason of each contradiction, a more detailed analysis is required. Here, two properties of the assignment table obtained from the reformulated overall constraint can be exploited.

Note that for simplicity we always refer to the assignment table in the presentation. As shown later in the implementation the assignment table needs not to be build explicitly.

PROPERTY 3.1 *The value of variable s_i is 0 for all solutions (i.e. in each row of the table) iff the respective constraint C_i is self-contradictory, that is C_i has no solution.*

Proof: $\Rightarrow$: We show this by contraposition: If C_i has at least one solution, then there is a row where s_i is 1. Obviously this solution (row) can be constructed by assigning 1 to s_i and 0 to s_j for $j \neq i$, because $(s_i \rightarrow C_i) = \overline{s}_i \vee C_i = 0 \vee C_i = C_i = 1$ and $(s_j \rightarrow C_j) = \overline{s}_j \vee C_j = 1 \vee C_j = 1$ for $j \neq i$.
$\Leftarrow$: To satisfy C' each element of the conjunction must evaluate to 1, so $(s_i \rightarrow C_i) = \overline{s}_i \vee C_i$. Since C_i has no solution (C_i is always 0) s_i must be 0. ■

Thus, each constraint C_i whose s_i variable is always assigned to 0, is a reason for the contradictory overall constraint C.

PROPERTY 3.2 *The value of variable s_i is don't care for all solutions (i.e. for all rows of the table) if the constraint C_i is never part of a contradiction of C.*

Proof: $\Rightarrow$: This property is shown by contradiction. Assume that s_i is don't care for all solutions and C_i is part of a contradiction. Then, without loss of generality there has to be another satisfiable constraint C_j such that $C_i \wedge C_j = 0$.[5] If s_j is set to 1 and all other constraints C_k with $k \neq j$ are disabled by $s_k = 0$,

[5] According to Property 3.1 both constraints C_i and C_j have at least one solution.

then C' is 1. However, switching s_i to 1 is not possible due to the conflict of C_i and C_j. But this contradicts the assumption that the value of s_i is don't care for all solutions.
$\Leftarrow$: Because the constraint C_i is never part of a contradiction, C_i can be enabled or can be disabled. In other words, s_i can be set to 0 and also to 1 for each solution of the overall constraint, which is equivalent to s_i being don't care. ■

Thus, each constraint C_i whose s_i variable is always don't care, is not part of a reason for the contradictory overall constraint. Therefore these constraints are not presented to the verification engineer and can be left out in the next steps.

EXAMPLE 3.13 *Consider again Example 3.12. Because the value of s_0 is 0 for all solutions, C_0 is self-contradictory. Thus, $R_0 = \{C_0\}$ is a reason for C. Since the value of s_1 is always don't care, C_1 is never part of a contradiction. As a result the first two constraints can be ignored in the further analysis.*

Note that the overall constraint of the example in Figure 3.11 has been specified to demonstrate the two properties. In practice, the number of constraints that are never part of a contradiction is considerably larger. Thus, applying Property 3.2 reduces the debugging effort significantly because every "non-relevant" constraint does not have to be considered any further by the verification engineer.

In fact, all *remaining constraints* (if there are any) are part of at least one contradiction. Furthermore, since contradictory constraints have been filtered out by Property 3.1 only a conjunction of two or more constraints causes a contradiction. Now the question is, how to identify the minimal contradicting conjunctions of the remaining constraints, i.e. the reasons?

EXAMPLE 3.14 *Again Example 3.12 is considered. The constraints C_0 and C_1 have been handled already according to Property 3.1 and Property 3.2. Now, the conjunction of two or more of the remaining constraints, C_2, C_3, C_4, C_5 and C_6, causes a contradiction. Only identifying the complete product of all these constraints certainly does not help to resolve the conflict. In fact, each individual contradiction has to be presented to the verification engineer. But this requires the computation of all minimal reasons according to Definition 3.10. In the example, three reasons can be found in total: $R_1 = \{C_2, C_4\}$ and $R_2 = \{C_3, C_4\}$ which overlap as well as $R_3 = \{C_5, C_6\}$ which is independent of R_1 and R_2.*

To find the minimal reason for each contradiction, all constraint combinations are tested for a contradiction starting with the smallest conjunction. For each tested combination the respective s_i variables are set to 1. Thus, if the conjunction $C_{i_1} \wedge \cdots \wedge C_{i_k}$ leads to a contradiction ($(s_{i_1} = 1) \wedge \cdots \wedge (s_{i_k} = 1) \wedge C' \equiv 0$), then this combination is a reason of C. Minimality is ensured by building the constraint combinations in ascending order with respect to their size and skipping each superset of a previously found reason. Since the overall problem has already been simplified by exploiting Property 3.1 and Property 3.2, the described procedure has to be applied only for a small set of constraints, i.e. the remaining ones. This is the key to the efficiency of the overall contradiction analysis procedure.

The next section presents the details on the implementation of the overall contradiction analysis approach.

3.3.2 Implementation

As explained in Section 3.1.2, the SCV library uses BDDs for the representation of constraints. More precisely the characteristic function of the overall constraint is represented as a BDD. This characteristic function is true for all solutions of the constraint, false otherwise. We implemented the contradiction analysis approach using the SCV library. Therefore our implementation is "BDD driven".

The pseudo-code of the contradiction analysis approach is shown in Algorithm 3.1. As input the approach starts with the BDD representation of the reformulated constraint C' and the set of all constraint variables V. At first, all constraint variables are existentially quantified from the reformulated constraint (line 2). Thus, the resulting function C'' only depends on the s_i variables. In other words, this function is the symbolic representation of the assignment table described in the previous section. In general the quantified BDD is much more compact than the BDD for the reformulated constraint. Thus, the following BDD operations can be executed very fast.

After quantification the two sets $\mathcal{R}$ and S are initialized to be empty. $\mathcal{R}$ stores all reasons that are found. Note that for simplicity $\mathcal{R}$ contains the sets of the corresponding s_i variables of a reason, not the constraints themselves. The set S is used to save all s_i variables that are passed to the detailed analysis later. So this set corresponds to the remaining constraints. Then, for each constraint C_i it is checked whether C_i is either contradictory (line 8) or never part of a contradiction (line 11) according to Property 3.1 and Property 3.2. In the former case the respective s_i variable is added to the set of reasons $\mathcal{R}$ (line 10). Both checks are conducted on the quantified representation C'' of the reformulated constraint, that is

- To check whether s_i is 0 for all solutions (see Property 3.1) the conjunction $C'' \wedge s_i = 1$ is carried out. If the result is the constant zero-function, s_i is never 1 in any solution, i.e. s_i is always 0. Thus, C_i becomes a reason.

Algorithm 3.1: contradictionAnalysis(BDD C', set of constraint vars V)

Result: Set $\mathcal{R}$ of reasons

1 `// abstraction`
2 $C'' = \exists v_1, \ldots, \exists v_{|V|}\ C'$
3 `// initialization`
4 $\mathcal{R} = \emptyset$; `// reasons of contradictions`
5 $S = \emptyset$; `//` s_i `variables for detailed analysis`
6 `// test properties`
7 **for** $(i = 0 \ldots n-1)$ **do**
8   **if** $((C'' \wedge s_i = 1) \equiv 0)$ **then**
9   `//` C_i `is self-contradictory`
10         $\mathcal{R} = \mathcal{R} \cup \{\{s_i\}\}$;
11   **else if** $((C'' \wedge s_i = 0) \equiv (C'' \wedge s_i = 1))$ **then**
12   `//` C_i `is not responsible for over-constraining`
13   **else**
14   `//` C_i `is selected for detailed analysis`
15         $S = S \cup \{s_i\}$;
16 `// detailed analysis`
17 **foreach** $(X \in \mathcal{P}(S))$ **do**
18   `// from the smallest to the largest`
19   **if** $(\exists X' \in \mathcal{R} : X' \subset X)$ **then**
20   `// ensure minimality`
21   **continue** ;
22   **if** $((C'' \wedge \bigwedge_{s_i \in X} s_i = 1) \equiv 0)$ **then**
23   `// subset leads to over-constraining of` C
24         $\mathcal{R} = \mathcal{R} \cup \{X\}$;
25 **return** $\mathcal{R}$;

- The check whether s_i is don't care in all solutions (see Property 3.2) is carried out by $(C'' \wedge s_i = 0) \equiv (C'' \wedge s_i = 1)$. If the respective BDDs are equal, it has been shown that s_i is don't care, since regardless of the value of s_i the solutions are identical. Therefore, the constraint C_i is not relevant for a contradiction and thus neither added to the set $\mathcal{R}$ nor to the set $\mathcal{S}$.

If both properties cannot be applied (line 13), then the respective constraint C_i is part of a contradiction caused by the conjunction of C_i with one or more other constraints. Thus, C_i is passed to the detailed analysis by inserting the respective s_i into S (line 15).

Finally, the detailed analysis for all elements in S – the remaining constraints – is performed (line 17–24). First, the power set $\mathcal{P}(S)$ of S is created resulting in all subsets, i.e. combinations, of constraints considered for detailed analysis. Note that we exclude the empty set as well as all sets which only contain one element from the power set, since this is already covered by Property 3.1. Furthermore, during the construction the elements of the power set are ordered according to their cardinality. Then, for each subset X, i.e. for each combination, the conjunction of the respective constraints is tested for a contradiction. Therefore, the conjunction of the current combination X – represented as a cube of all variables $s_i \in X$ – and C'' is created, i.e. all respective constraints C_i are enabled (line 22). If the conjunction leads to a contradiction, then X is a reason and thus, X is added to $\mathcal{R}$ (line 24). To ensure minimality each contradiction test of a subset X is only carried out if no reason $X' \in R$ exists such that $X' \subset X$ (line 19–21), i.e. no subset of X has already been identified as reason for a contradiction (see also Definition 3.10).

In summary, the presented contradiction analysis procedure computes all minimal reasons $\mathcal{R}$ of a contradictory overall constraint C. First, the proposed reformulation of the overall constraint allows a representation where all contradictory constraints can be disabled. From this representation a much more compact one is computed by quantification. All following operations have to be carried out on this representation only. Then, the two properties are applied which significantly reduces the problem size since only $2^{n-|Z|-|DC|}$ instead of all 2^n subsets have to be considered in the detailed analysis (Z denotes the set of self-contradictory constraints, and DC denotes the set of constraints that are not part of a contradiction). In practice, especially the number of "non-relevant" constraints that belong to the set DC is very large, so the input for the detailed analysis shrinks considerably.

3.3.3 Experimental Results

This section provides experimental results for the contradiction analysis. First, different types of contradictions are discussed that have been observed in practice. Then, we show the efficiency of our approach using several testcases. Finally, we demonstrate the advantages of our approach in an industrial setting. We briefly discuss a constraint-based simulation environment used at AMD DDC for verification of SoC designs. By means of a concrete example we will show how time spent on debugging constraint contradictions is significantly reduced by our approach.

In all examples the partitioning of the constraints is given according to the specification in the constraint classes, i.e. each C_i in the following corresponds to a separate SCV_CONSTRAINT() macro (see also Section 3.3.1). The contradiction analysis is started by an additional command line switch and runs fully automatic in the SCV library environment.

Types of Contradictions

We have identified different types of contradictions. In the following the general structure is shown by means of examples. We assume that self-contradictory constraints as well as "non-relevant" constraints have been removed. Assume k constraints are left. Then, one of the following cases are possible which are identified fully automatic by our approach:

1. There is exactly one contradiction that is caused by all k constraints. Here, no other subset of the constraints forms a contradiction and thus all constraints are the only reason for the over-constraining. A simple and a more complex example is shown in Figure 3.12(a).[6]

2. There are at least two contradictions. This case can be refined further:

   (a) Our approach determines m disjoint partitions from the constraint set. This means our approach has identified m independent contradictions. An example is given in Figure 3.12(b). In this example for the constraint set $\{C_0, C_1, C_2, C_3\}$ the two reasons $R_0 = \{C_0, C_1\}$ and $R_1 = \{C_2, C_3\}$ are determined.

   (b) There is at least one overlapping, i.e. at least one constraint C_i is part of at least two reasons. Also here an example is given in Figure 3.12(c). This example shows the two reasons $R_0 = \{C_0, C_2, C_4\}$ and $R_1 = \{C_1, C_3, C_4\}$. Obviously C_4 is part of both reasons.

Our proposed approach is able to identify the minimal reason for all these types of contradictions.

$C_0 \Leftrightarrow$ a() == 6
$C_1 \Leftrightarrow$ a() == 10

$C_0 \Leftrightarrow$ a() == 6
$C_1 \Leftrightarrow$ a() == 5 || b() == 1
$C_2 \Leftrightarrow$ b() == 2 || a() < 2

(a) Exactly one contradiction caused by all constraints

$C_0 \Leftrightarrow$ a()==1
$C_1 \Leftrightarrow$ a()>2
$C_2 \Leftrightarrow$ b()==1
$C_3 \Leftrightarrow$ b()==2

(b) Independent contradictions

$C_0 \Leftrightarrow$ a() == 6
$C_1 \Leftrightarrow$ b() == 7
$C_2 \Leftrightarrow$ a() < 3 || c() == 1
$C_3 \Leftrightarrow$ b() < 3 || d() == 1
$C_4 \Leftrightarrow$ c() + d() == 0

(c) Overlapping contradictions

Figure 3.12. Types of contradictions

[6]The reasons are marked by brackets.

Effect of Property 1 and Property 2

Applying the two properties introduced in Section 3.3.1 significantly reduces the complexity of the contradiction analysis since each matched constraint can be excluded from further considerations. To show the increasing efficiency we tested our approach for several examples which contain some typical over-constraining errors (e.g. typos, contradicting implications, hierarchical contradictions, etc.).

For the considered constraints we give some statistics in Table 3.3. In the first column a number to identify the testcase is given. Then, in the next columns information on the constraint variables and their respective sizes are provided. Finally, the total number of constraints is given. The results after application of our contradiction analysis are shown in Table 3.4. The first four columns give some information about the testcase, i.e. the number of constraints in total (n), the number of contradictions/reasons ($|\mathcal{R}|$), and the run-time in CPU seconds needed to construct the BDD in the SCV library (BDD TIME). The next columns provide the results for the analysis approach without (W/O PROPERTIES) and with the application of the properties (WITH PROPERTIES), respectively. Here the number of checks in the worst case (2^n or $2^{n'}$, respectively), the number of checks actually executed by the approach ($\#\surd$), and the run-time for the detailed analysis (TIME) are given. Additionally, the number of "non-relevant" constraints ($|DC|$) and self-contradictory constraints ($|Z|$) obtained by the two properties are provided.

Table 3.3. Constraint characteristics

#	Boolean	Int	Long	Bits	Constraint (n)
1	10	8	–	328	15
2	3	3	6	483	16
3	10	10	–	330	26
4	8	40	–	1,288	50
5	5	30	15	1,925	53

Table 3.4. Effect of using Property 1 and Property 2

			BDD	W/O PROPERTIES			WITH PROPERTIES				
#	n	$\|\mathcal{R}\|$	TIME	2^n	$\#\surd$	TIME	$\|Z\|$	$\|DC\|$	$2^{n'}$	$\#\surd$	TIME
1	15	1	5.48	32,768	24,577	4.12	0	13	4	4	0.06
2	16	3	14.90	65,536	26,883	11.25	1	8	128	107	0.04
3	26	1	22.30	67,108,864	–	TO	0	21	32	32	0.30
4	50	3	35.96	$> 1.1 \cdot 10^{15}$	–	TO	0	42	256	190	2.10
5	53	2	238.07	$> 9.0 \cdot 10^{15}$	–	TO	0	47	64	55	9.77

The results clearly show, that identifying all reasons without applying the properties leads to a large number of checks in the worst case (e.g. $2^{53} \geq 9.0 \cdot 10^{15}$ in example #5). Since the detailed analysis is only carried out for a combination X if no subset of X has already been identified as reason, the real number of checks is smaller. In contrast, when the properties are applied most of the constraints can be excluded for the analysis since they are "non-relevant". This significantly reduces the number of checks to be performed at detailed analysis. Instead of all 2^n only $2^{n-|Z|-|DC|}$ checks are needed in the worst case (only 64 in example #5). As a result the run-time of the detailed analysis is magnitudes faster when the properties are applied. Moreover, for the last three testcases the reasons can be determined within the timeout of 7,200 CPU seconds only when the properties are applied.

Real-Life Example

The constraint contradiction analysis algorithm has been evaluated using a real-life design example. The used verification environment is depicted in Figure 3.13 (see also [GSD06]).

The *Design Under Verification* (DUV) is a PCIe root complex design with an AMD-proprietary host bus interface which is employed in a SoC recently developed by AMD. The root complex supports a number of PCIe links. The verification tasks are to show (1) that transactions are routed correctly from the host bus to one of the PCIe links and vice versa, (2) that the PCIe protocol is not violated and (3) that no deadlocks occur when multiple PCIe links communicate to the host bus at the same time.

Host bus and PCIe links (only one depicted in Figure 3.13) are driven by *Bus Functional Models* (BFMs) which convert abstract bus transactions into the detailed signal wigglings on those buses. The abstract bus transactions are generated by means of random generators (denoted by *G*) which are in turn controlled by constraints. Bus monitors observe the transactions sent into or from either interface and send them to checkers which perform the end-to-end transaction checking of the DUV. The verification environment is implemented in SystemC 2.1, the SCV library and SystemVerilog, with a special co-simulation interface synchronizing the SystemVerilog and SystemC simulation kernels. The constraint-based verification methodology was chosen in order to both reduce effort in stimulus pattern development and to get high coverage of stimulation corner cases. The PCIe and host bus protocol rules were captured in SCV constraint descriptions and are used to generate the contents of the abstract bus transactions driving the BFMs.

The PCIe constraint used to control stimulus generation within the PCIe transaction generator is a layered constraint. The lower level layer describes generic PCIe protocol rules and is comprised of a number of 16 constraint

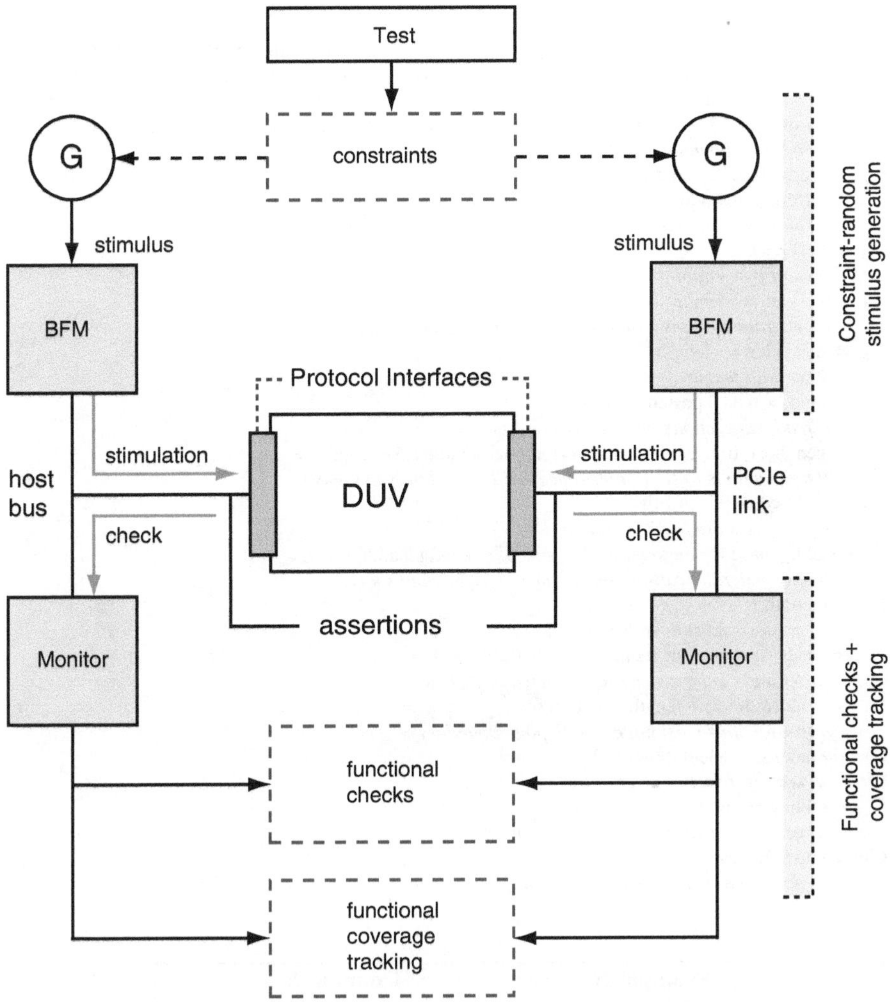

Figure 3.13. Architecture for verification at AMD DDC

expressions. They are shown in Figure 3.14(a) (denoted from C_0 to C_{15}).[7] The meaning of the constraint variables is given in Table 3.5. The upper level constraint layer imposes user-specific constraints on the generic PCIe constraints (denoted by C_{U_i}) in order to generate specific stimulus scenarios. Generic PCIe constraints and user-defined constraints are usually

[7]Bit operators are used as introduced for the SCV library in Section 3.2.1. For example, the constraint C_{12} makes use of the "bitwise AND".

$C_0 \Leftrightarrow$ (addr_space != memory ||
((mem_addr_base0 <= addr) && ((addr+length) <= mem_addr_base0 + mem_size0)))
|| ((mem_addr_base1 <= addr) && ((addr+length) <= mem_addr_base1 + mem_size1)))
// address boundaries for memory
$C_1 \Leftrightarrow$ (addr_space != io || ((io_addr_base <= addr) && ((addr + length)<= io_addr_base + io_size)))
// address boundaries for io
$C_2 \Leftrightarrow$ (addr_space != config ||
((cfg_base_addr <= addr)&&((addr+length) <= config_base_addr + config_size)))
// address boundaries for config
$C_3 \Leftrightarrow$ be[] <= 0xf
// valid byte enables are in 0x0...0xf
$C_4 \Leftrightarrow$ be[].len == length
// generate as many byte enables as we have dword data
$C_5 \Leftrightarrow$ data[].len == length
// set data length
$C_6 \Leftrightarrow$ cmd != read || posted == false
// read transactions are always non−posted
$C_7 \Leftrightarrow$ gen_host_trans.addr_space == memory || (addr&3)+length <= 4
// transactions to IO/Config space are 1 dword (4bytes) only
$C_8 \Leftrightarrow$ addr <= 0xFFFFFFFF
// addresses are in 32 bit range
$C_9 \Leftrightarrow$ addr_space == memory || addr_space == config || addr_space == io
// only generate transactions in memory/IO/config space
$C_{10} \Leftrightarrow$ length > 0
// requests must have length > 0
$C_{11} \Leftrightarrow$ addr_space == sr::mem || addr <= 0xFFFFFFFF
// IO and config space are restricted to 32 bits
$C_{12} \Leftrightarrow$ (addr&4095) + length <= 4096
// transactions must not cross 4k page boundary
$C_{13} \Leftrightarrow$ (addr&3) + length <= 128
// keep transaction length to max. 128 bytes
$C_{14} \Leftrightarrow$ tkind == request
// generate requests only (not responses)
$C_{15} \Leftrightarrow$ msr == false
// do not generate MSR accesses

(a)

Example 1:	**Example 2:**
$C_{U_1} \Leftrightarrow$ length > 128	$C_{U_2} \Leftrightarrow$ addr == 4000
	$C_{U_3} \Leftrightarrow$ length == 100

(b)

Figure 3.14. PCIe transaction generator constraint with examples

developed by different verification engineers; the former by the designer of the test environment and the latter by the engineer who implements and runs the tests.

The engineer writing the tests and hence the user-specific constraints which are layered on top of the generic PCIe constraints is faced with the problem to resolve contradictions which are generated by imposing the user-defined

Table 3.5. Definition of random variables used in the PCIe constraint

Variable name	Description
addr	Transaction address (64 bits)
addr_space	Transaction address space (memory,io,config)
tkind	Transaction kind (request,response)
cmd	Transaction command (read,write)
msr	Transaction is targeted at MSR space
posted	Transaction is posted (yes/no)
length	Transaction size in dwords
be[]	Array of byte enables (one per each dword data)
data[]	Array of dword (32 bit) data
be[].len	Length of byte enable array
data[].len	Length of data array
[io|mem|cfg]_addr_base0,1	io, memory and config space window base addresses
[io|mem|cfg]_size0,1	io, memory and config space window sizes

constraints on the PCIe generic constraints. Given the complexity of the constraints, this is usually a non-trivial task. Two real-life examples of contradictions that are not easy to resolve by manual constraint inspection are depicted in Figure 3.14(b).

In the first example the user sets the maximum transaction length to a value greater than 128 bytes (C_{U_1}), thereby causing a contradiction to constraint C_{13}, which states that the total transaction length must not exceed 128 bytes. In the second example, the user independently constrains the transaction address to byte address 4,000 (C_{U_2}) and the transaction length to 100 bytes (C_{U_3}). While both values, viewed independently, are each perfectly legal (the address should be in 32 bit range and the transaction length is less than 128 bytes), an over-constraining occurs. The reason identified by our approach is $R_1 = \{C_{12}, C_{U_2}, C_{U_3}\}$. By manual constraint inspection it is not immediately obvious that a PCIe protocol rule is violated when combining constraints C_{U_2} and C_{U_3}. However, reason R_1 found for the contradiction by our algorithm shows that when combining constraints C_{U_2} and C_{U_3}, then PCIe protocol rule C_{12} is violated: "A transaction must not cross a 4k page boundary". Our user constraints of transaction start address set to 4,000 and transaction length of 100 bytes would result in addresses that cross a 4 k page and therefore violate this constraint.

The presented algorithm was able to identify exactly the violating constraint expressions for both examples in about 30 CPU seconds fully automatic. The PCIe constraint to be analyzed contained a total of 21 random variables to be solved which are constrained by 17 and 18 constraint expressions for the respective examples. The total bit count for the random variables amounted to

781 bits. Without such an analysis capability, we would have had to spend a long time on manual constraint inspection in order to identify the root cause for the constraint contradiction. Thus, a significant speed-up of the contradiction debug cycle can be achieved.

3.4 Measuring the Quality of Testbenches

As presented in the previous section for system-level verification constraint-based simulation has been used. This allows to create high quality verification scenarios by constraints. However, even such a sophisticated verification technique does not include a measure how thorough the design was executed during the simulation. As the size of the testbench – given as directed tests or constraints – grows, the designer needs a reliable feedback about its quality.

In this section, an approach to measure the quality of the testbench is presented. Our analysis is based on dedicated code coverage techniques. They have been developed for SystemC models. By exploiting automated code instrumentation based on a SystemC front-end, for each test run a coverage report is generated that presents to the user all statements in the model that have not been executed during simulation. The report is based on the analysis of the exercised control flow statements. It includes exact source code references to unexecuted code blocks in combination with SystemC specific information, like process context and hierarchy information.

In software testing code coverage techniques have been used to measure the fraction of code that has been exercised by a test case [Bei90]. From this domain coverage methods have been derived and extended for HDLs. For Verilog or VHDL several approaches and tools exists (for an overview see e.g. [TK01]). However, no code coverage method to measure the quality of a SystemC testbench has been proposed. Note, that approaches based on standard C++ coverage tools (like e.g. the GNU coverage tool gcov [gco]) have several drawbacks. On the one hand the SystemC kernel is also included in the coverage analysis. On the other hand SystemC specific data, like e.g. context information or hierarchy information, is only implicitly available and has to be extracted manually. In the following we present an approach to overcome these limitations.

This section is structured as follows. In Section 3.4.1 we present our approach. We start with the description of the overall flow in Section 3.4.1. Then, we continue with a detailed description of the three phases of our approach in Section 3.4.2. Along the way we provide an example to show the effects of each phase. Experimental results for two SystemC designs are presented in Section 3.4.3. The first design is a RISC CPU and the second design is a TLM-based video processor.

3.4.1 Code Coverage-Based Approach

In this section the code coverage-based approach for measuring the quality of the testbench is introduced. Our approach consists of three phases: SystemC analysis, code instrumentation and coverage analysis. Before the details on the three phases are given the overall flow is presented. Throughout the description of the phases a simple example is used to demonstrate the effects of each phase.

Overall Flow

The overall flow of our approach is depicted in Figure 3.15. In the analysis phase the SystemC code of the DUV is parsed, analyzed and transformed into an *Abstract Syntax Tree* (AST) representation. This AST is traversed in the consecutive code instrumentation phase. During the traversal the original SystemC DUV is augmented with SystemC specific code that enables the collection of coverage information during simulation. Then, the rewritten SystemC DUV, the coverage library of our approach and the SystemC libraries are compiled into an executable. By running this executable, simulation is performed and the data structures available through our coverage library are filled. Finally, in the coverage analysis phase the collected data is interpreted and the coverage report is generated. By the report the verification engineer is informed which statements have not been executed using the tests defined in the testbench. This information is presented with exact source code references to unexecuted blocks in the original SystemC DUV including hierarchy. Furthermore the frequency of the execution of statement blocks can be given for further analysis.

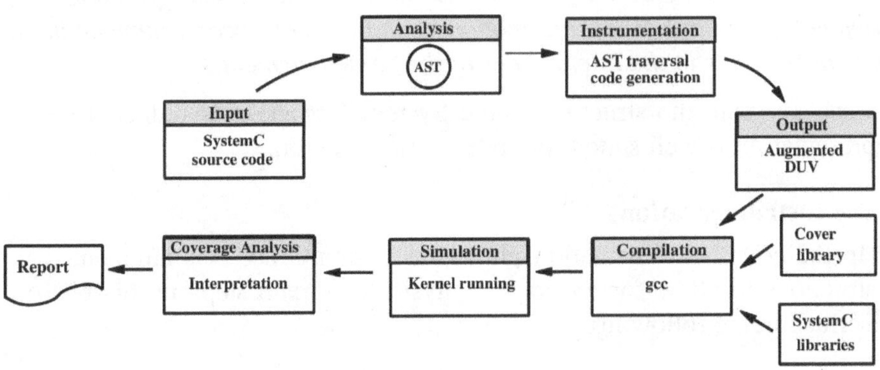

Figure 3.15. Overall flow of code-coverage based approach

3.4.2 Phases of Code Coverage-Based Approach

In the following we describe the three phases in more detail.

SystemC Analysis

For the transformation of the SystemC DUV into an AST the front-end from [FGC+04, GD06] is used. The important features of the front-end that are required for the proposed approach are briefly described in the following. The parser of the front-end was built with PCCTS (Purdue Compiler Construction Tool Set) [Par97]. PCCTS enables the description of the SystemC syntax in form of a grammar, provides facilities for AST construction and finally generates a parser. Note that the front-end has an exact source code reference including character positions of each token. Therefore, a special C++ pre-processor has been implemented to allow for identification of the SystemC macros before they are expanded. The correct source code information annotated to each node in the AST is very important for our approach. Without this information only a non-reliable feedback for the verification engineer would be possible. In the following the analysis phase is demonstrated by an example.

EXAMPLE 3.15 *Since we use a program counter of a RISC CPU also as example for the other phases, we give some details on this module. In order to address the 2048 entries of the program memory, the PC has an 11 bit register which holds the address of the current instruction. Output* pcout *holds this address. The output* pcinc *gives the address increased by one. An address can be loaded into the PC via the input* din, *if the input* le *(load enable) is set to 1. Using the* reset *signal, the PC can be set to 0. On every positive edge of the* clock *signal the current address is increased if the input* en *(enable) is set to 1. In Figure 3.16 the method that computes the* `next_state` *of the PC is shown. This method is sensitive to the positive clock. The internal register of the PC module is* `pc`. *Figure 3.17 depicts a sample of the AST of this method, which has been generated by our tool. Please note that for each AST node only a fragment of the available information is shown. The second number in each line corresponds to the line number of the parsed element.*

As can be seen, the structure of the SystemC program is reflected and this representation is well suited for code instrumentation.

Code Instrumentation

In the code instrumentation phase the SystemC DUV is augmented with instructions to allow for coverage analysis. The main steps in this phase are described in the following.

Coverage Library. First, the global variable `cov` is defined that holds an instance of our coverage class `COVER`. This class provides data structures like

```
void prog_count::next_state() {
 if (reset.read()) {
  pc = 0; //reset to adress 0
 } else {
   if(en.read()) {
     if(le.read()) {
       pc = din; //load address
     } else {
       // increase counter
       pc = pc.read() + 1;
     }
   } else {
     pc = pc.read();
   }
 }
}
```

Figure 3.16. Parts of the original SystemC DUV

```
10 IF
10  LPAREN
10  ID == "reset"
10  DOT
10      ID == "read"
10  LPAREN
10      RPAREN
10  RPAREN
10  LCURLY
11      ASSIGNEQUAL
11         ID == "pc"
11         OCTALINT
11      SEMICOLON
12      RCURLY
12  ELSE
12  LCURLY
13    IF
13     LPAREN
13        ID == "en"
...
```

Figure 3.17. AST of next_state method

hash tables for coverage statistics as well as wrapper functions to take care of the control flow inside the methods of the DUV. Furthermore, the class has methods to analyze the collected coverage data and to generate the report for the user.

AST Traversal and Code Instrumentation. While traversing the AST, first the member functions that belong to a SystemC module are identified. Then, in each function the conditions of the control flow statements are substituted with wrapper functions. The idea is to perform a call-back during the simulation and thereby notifying the coverage class which control branch has been taken. The following control statements are distinguished: IF, IF/ELSE, SWITCH-CASE, FOR-loop, WHILE-loop. Next, the wrapper functions are explained.

Wrapper Functions. For the IF, IF-ELSE, FOR-loop and WHILE-loop the condition of the control statement is replaced by a wrapper function call. The arguments of the wrapper functions are

1. The condition of the control statement (as Boolean and string).
2. The type of control statement.
3. Start position and end position of the block(s) that are executed if the control condition evaluates to true/false.
4. File name of the current method.
5. Class name if available.
6. Current method name.
7. `this` pointer, in case of a member function. The `this` pointer is used to distinguish between several instances of the same module.

The following example demonstrates the application of a wrapper function for an IF-ELSE control statement.

EXAMPLE 3.16 *Consider again the program counter in Figure 3.16 and focus on the if statement in line 2 and the corresponding else-branch starting in line 4. The condition of the if statement is the expression* `reset.read()`. *This expression is replaced by the function* `wrapperStatement(...)`. *The instrumented code is depicted in Figure 3.18. The first and second argument of this function hold the condition as a Boolean and as a string, respectively. The third argument reflects the type of the condition statement – here* `tIFELSE`. *Then, the next four numbers mark the if-block, i.e. the if-block starts in line* `10` *at the absolute character position* `125` *and ends in line* `12` *at character position* `203`. *The next two numbers give the same information for the else-block, but only the end position of the else-block is used; the else-block ends in line* `22` *at character position* `419`. *Then, the file name where the method is implemented (*`prog_count.cc`*), the class name (*`prog_count`*), the method name (*`next_state`*) and the* `this` *pointer are given.*

```
#include "cover.h"
#include "label.h"
extern COVER *cov;

#include"prog_count.h"
...
void prog_count::next_state() {
if(cov->WrapperStatement(reset.read(),"reset.read()",
    tIFELSE,10,125,12,203,22,419,"prog_count.cc", "
    prog_count", "next_state",this)) {
  pc = 0;
} else {
...
```

Figure 3.18. Instrumented code of the next_state method

In a SWITCH-CASE statement at the beginning of each `case` block we instrument a wrapper function that has as additional argument the value of the current case. After a SWITCH-CASE statement a wrapper function is instrumented that enables the propagation of all possible CASE values.

Note that the approach is able to handle also nested variants of all types of control statements. In the next section the coverage analysis phase is explained.

Coverage Analysis

After the compilation of the instrumented SystemC code the coverage analysis is executed during simulation. Based on the instrumented wrapper functions the instance of the cover class collects all the coverage data. The main data structures in the cover class are based on *Standard Template Library* (STL) maps. As unique keys the arguments of the wrapper functions are transformed into a string representation. To each coverage point we associate two counters to track the frequency of the evaluation of the corresponding condition to true or false. For case statements obviously only one counter is needed. Finally, in the coverage report that is started by a call from `sc_main` after the end of the simulation, the coverage data is analyzed. For IF, IF/ELSE a warning is generated if the condition was always true/false and thus a block was never executed. In case of FOR-loops or WHILE-loops the user is informed if the condition was false all the time and therefore the loop body was skipped. For SWITCH-CASE statements each case is identified that was never activated. In total this allows to argue about the quality of the tests defined by the testbench. If blocks have been identified that have never been executed, these blocks are dead code or the testbench has to be improved.

In the following example the results of the coverage analysis are shown for the program counter.

```
<< COVERAGE REPORT >>

IF-ELSE Statement: *IF-BLOCK
    NOT EXECUTED*
  File name: prog_count.cc
  Class: prog_count
  Instance: pc
  Func. Member: next_state
  Condition: le.read()
  IF start: line 14 pos 246
  IF end:   line 16 pos 322
  count total: 87
  count TRUE: 0 count FALSE: 87
```

Figure 3.19. Coverage report for program counter

EXAMPLE 3.17 *A directed testbench has been written for the program counter shown in Figure 3.16. The testbench includes three tests. We applied our approach for this example. The automatically generated coverage report is shown in Figure 3.19. As can be seen the scenario to load a value into the program counter by setting load enable to 1 was not executed. We added another test for this behavior and thereby closed this gap.*

3.4.3 Experimental Results

In this section we apply the approach to two examples. The first example is a hardware oriented model, a RISC CPU is considered. The second example is a system for color region recognition in video data.

Hardware Model: RISC CPU

Before we apply our method to the RISC CPU the basic data of the CPU is briefly reviewed (see [GKG+05, Kue06] and Section 5.2.3 for more details).

Specification. In Figure 3.20 the components of the RISC CPU are shown. The CPU has been designed as a Harvard architecture. The data width of the program memory and the data memory is 16 bit. The size of the program memory is 4 kByte and the size of the data memory is 128 kByte. The length of an instruction is 16 bit. We briefly describe the five different classes of instructions in the following: 6 load/store instructions, 8 arithmetic instructions, 8 logic instructions, 5 jump instructions and 5 other instructions. For the RISC CPU a compiler has been implemented which generates object code from an assembler program. This object code runs on the SystemC model, i.e. the model of the CPU executes an assembler program.

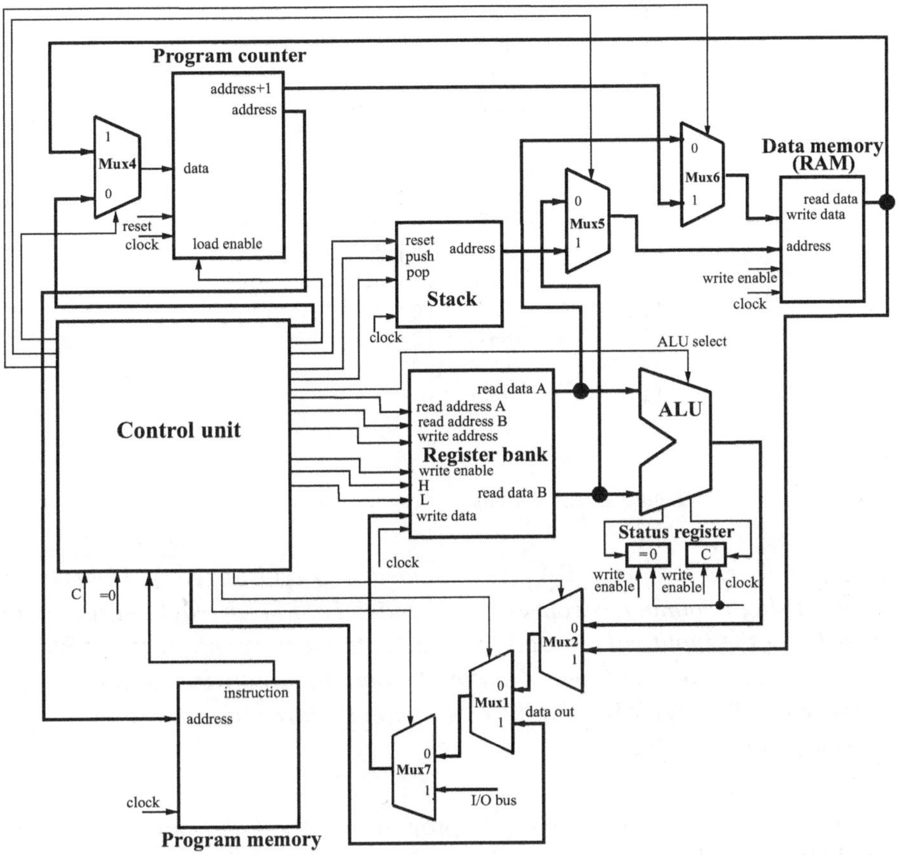

Figure 3.20. RISC CPU including memories and full data paths

Testbench Quality. Based on successful simulation of each component the designer starts with the simulation at the top level. (More details on top-level verification is given later). For this purpose usually a high-level testbench is created that enables a black-box test of the design. For the CPU such a testbench corresponds to the execution of a set of assembler programs including the analysis of the simulation results. In the following we describe how the high-level testbench was created and how this process was improved by our approach. The SystemC model of the RISC CPU was automatically instrumented with code to analyze coverage. The following non-trivial assembler program was formulated to test the CPU.

EXAMPLE 3.18 *The assembler program shown in Figure 3.21 converts a set of numbers into gray-code. The gray-code encodes numbers such that in the binary encoding adjacent numbers have a hamming distance of 1. The number n of elements to be converted is given in the data memory at address 0. After*

```
          LDL R[6], 0
          LDH R[6], 0
          LDL R[2], 0
          LDH R[2], 0
          LDD R[3],R[2]
loop1:
          ADD R[2],R[2],R[1]
          LDD R[4],R[2]
          ADD R[5],R[4],R[0]
          SHR R[5],R[5]
          XOR R[6],R[4],R[5]
          STO R[2], R[6]
          SUB R[3],R[3],R[1]
          JNZ loop1
          HLT
```

Figure 3.21. Assembler program for gray code

clearing registers R[6] *and* R[2], n *is loaded into register* R[3]. *Then, in the loop each single number is converted. The idea is to invert each bit if the next higher bit of the input value (read from the data memory into register* R[4]*) is set to one. Therefore the input is shifted by one and a bitwise XOR operation is performed. The result* R[6] *of the conversion is stored in the data memory at the same position as the input.*

After simulation of the gray-code program on the CPU our approach reported unexecuted code fragments in the following modules: `stack_point`, `mux4`, `mux5`, `mux6`, `mux7` and `alu`. The handling for the cases of push and pop operations in the `stack_point` module was not tested, since the inputs from the control unit to this module have been zero during the complete simulation. To test this behavior another program that uses push and pop instructions had to be added.

For the multiplexor modules we found that in the method `do_select` which describes the functionality of a multiplexor only the ELSE-block for the select condition was simulated. For the CPU this observation corresponds to the fact that the select inputs of the multiplexers have been zero all the time and thus only one data input was routed to the multiplexor output. As can be seen in Figure 3.20 all multiplexers belong to the data path of the CPU. To also test the effects on the CPU in case of data coming through the other input, a different data path has to be activated. The multiplexor `mux5` is part of the stack pointer data path and thus was tested by using stack pointer operations (see above). For `mux4` and `mux6` the alternative data path is activated by adding a program that uses sub-routine calls. For `mux7` we set the select input to one by an additional program that uses I/O instructions.

In case of the ALU several CASE statements of the main SWITCH statement have not been executed since not all operations of the ALU are activated by the considered assembler programs. Therefore we created another program to check the remaining arithmetic operations.

In total by adding additional assembler programs to the testbench the quality of the testbench was improved. Here our approach supported the verification engineer by directly pointing to untested functionality of the RISC CPU.

High-Level Model: Color Region Recognition

In the second example we applied our approach to a high-level SystemC model of a video processor SoC. In contrast to the RISC CPU (which has been implemented as an RTL design), this model is a very abstract system-level design and uses TLM features intensively.

Specification. The configurable model *EmViD* consists of a set of SystemC cores that can be integrated to build a video processor. For video input and output, abstract TLM channels are used. The video processing IP cores use the *SystemC High-level Interface Protocol* (SHIP) [Kli05] for data exchange over these channels. Communication with the main memory (DDR RAM) is established by ST's TAC protocol [Mic05]. In the following, we consider a SoC for color region recognition that is based on EmViD cores. The system processes video frames in real-time and draws rectangles around detected regions. A high-level schematic of the system is shown in Figure 3.22. The system

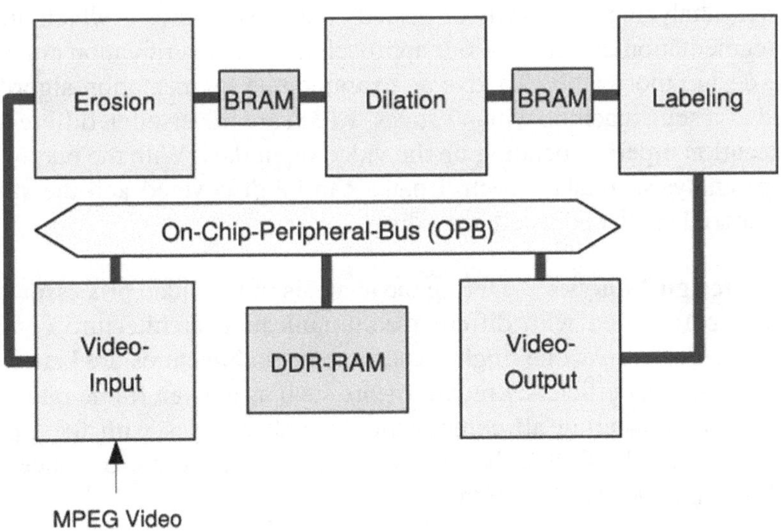

Figure 3.22. Color region recognition schematic

has been configured as a pipelined architecture and for the connection of the DDR RAM an IBM CoreConnect *On-Chip Peripheral Bus* (OPB) is used. The complete transaction-level interconnect (including an OPB simulation model) is set up using the GreenBus TLM fabric [KGB+06]. EmViD can be found at [EmV].

The video processing starts by reading in an MPEG video as video input. Then, erosion and dilation are performed. In the labeling stage the regions are recognized and the rectangles are added. Afterwards the core outputs the image to a display.

Testbench Quality. As a concrete application we decided to detect skins in the video data. We set the color range for the recognition accordingly. The system segments the processed video data in the labeling phase. Therefore adjacent pixels are analyzed and the image is partitioned into a set of regions using the defined color information.

In the overall video processor system the high-level testbench consists of the video data (coming from video files or a camera). We applied our approach to the system. We simulated the system with different video files and observed that depending on the video data different parts of the system have not been executed. For example, in the `morph_segm` module (labeling phase) the segmentation algorithm checks the minimum region size with an IF-condition. For video data that contains no skins or very small areas no regions are detected. Here, our approach presents directly the SystemC file with the exact source code position of the never executed block(s). Note that this improves the debugging during the development of such high-level models significantly. Moreover, analyzing the results of nested control structures – which are used in the segmentation algorithm – our approach helps the verification engineer to test the design thoroughly. To give an example, the segmentation algorithm is realized as a state machine with 47 states, which are traversed in different (partial) execution orders depending on the video input data. With the output of the coverage analysis, untaken control paths can be discovered and the stimulus video material can be adjusted accordingly.

Further Design Analysis. During the analysis of the video processor model, we also experimented with different communication architecture configurations for the design. As one might expect, some architectures are better suited than others to meet efficiency requirements such as a given frame rate. In particular, when connecting all components to a shared bus with fixed-priority scheduling (here, the OPB), the overall video processing performance highly depends on the priority allocation.

Table 3.6. Video processor execution traces

Config	# ex. video	# ex. detect.	FPS video	FPS detect.	Comment
Bus only model 1	500	500	24.98	24.98	Ascending priority
Bus only model 2	451	872	22.55	43.60	Higher detection priority
Mixed bus/pipeline model 1	500	500	24.98	24.98	Lower pipeline priority
Mixed bus/pipeline model 2	500	999	24.98	49.90	Higher pipeline priority

We utilized the ability of our coverage analysis to count the number of executions for the various processes in the model in order to identify the location of communication bottlenecks in design configurations with poor frame rates. Table 3.6 presents some results of the experiments.

The column *#ex video* shows the total number of video frames successfully sent from the video input component (MPEG decoder) to the video output component (display controller). The column *#ex detect.* shows the total number of video frames processed by the region detection. From these numbers the overall frame rates have been calculated (columns *FPS video* and *FPS detect.*). Row 1 and row 2 show the frame rates we got with a bus-only model. While in row 1, the bus access priorities were assigned in ascending order according to the sequence of video processing stages in the model, in row 2 we assigned a higher priority to the region detection components than to the video display data path. As expected, the frames per second processed for region detection goes up, but as an unintentional side effect due to higher bus workload, the number of video frames displayed per second drops down. Rows 3 and 4 show the results we achieved with a mixed bus/pipeline model as depicted in Figure 3.22. Here, we could considerably increase the video display frame rate by just swapping the bus access priorities of two components. With this setup, $\approx$25 frames per second full resolution live video display is achieved while the region detection runs at the high rate of $\approx$50 frames per second.

Overall, the presented approach allows to measure the quality of a SystemC testbench and hence helps to improve the verification quality. Moreover, the TLM example revealed that our analysis methodology can also support design space exploration by providing data on execution counts.

3.5 Summary and Future Work

In this chapter techniques for system-level verification have been presented. At first, constraint-based simulation in general and in the context of the SCV library has been described. This technique improves simulation by using formal methods for generating verification scenarios from constraints.

Then, two disadvantages of the SCV library have been resolved: First, new operators to simplify constraint specification have been integrated in the SCV library. The new operators are important especially for partially refined design descriptions. Second, the uniform distribution of all constraint solutions is guaranteed if constraint variables are fixed to a certain value. By this, the chance of entering unexplored regions of the design state space is maximized.

In the third part of the chapter a fully automated approach to analyze conflicts in contradictory constraints has been introduced. The method identifies all minimal reasons that are responsible for a contradiction in one single step. Thus, the manual debugging process is replaced by an automatic method. The approach together with improvements from above has been evaluated in a verification environment at AMD DDC. The experimental results have shown that the debugging time can be reduced significantly.

To ensure the resulting verification quality an approach to measure how thorough the design was tested by the testbench has been presented. The method uses dedicated code coverage techniques based on a SystemC front-end. Thus, the not tested design parts are presented to the user in form of a coverage report. Overall, the approach helps to enhance the testbench significantly since the manual quality check is removed.

Improving the underlying constraint-solving techniques of the SCV library remains a topic for future work. Promising directions are the integration of a *Satisfiability Modulo Theories* (SMT) solver for the stimuli generation. A first approach is presented in [WGHD09]. Driving the constraint-based simulation automatically into untested design parts by using the results of the coverage report is another interesting direction. A first step in this direction for pure combinational SystemC designs exploiting code coverage information has been proposed in [DCdS07].

In summary, the presented verification techniques in combination with the coverage check guarantees a high quality model at the system level. In the next chapter the verification at the block level is considered to ensure fully verified blocks before considering the entire top level where the verification of the communication between all synthesizable blocks has to be carried out.

Chapter 4

BLOCK-LEVEL VERIFICATION

In this chapter techniques for verification at the block level are presented. Figure 4.1 shows the respective parts of the proposed design and verification flow that are described in this chapter. The motivation for considering components of the system at the block level before addressing the top level in more detail is as follows. Based on the SystemC design methodology the system is stepwise refined and finally consists of hierarchical modules (with the respective functionality) and interfaces for communication. At this point all parts of the system are synthesizable. But from the verification perspective along the refinement process until reaching the synthesizable descriptions only simulation-based techniques have been used to check that the specification is met. Even with the strong constraint-based simulation methods and the complementing testbench quality check as presented in Chapter 3 typically not all design errors can be found. Thus, in the following formal methods are applied to the blocks of the design. Thereby, their functional correctness can be guaranteed.

The middle of Figure 4.1 shows the proposed verification techniques as well as the corresponding quality check for the block level. First, in this chapter a property checking approach for SystemC is presented. The approach uses the front-end of [FGC$^+$04], which is part of the SystemC design environment SyCE[1] [DFGG05], to generate a *Finite State Machine* (FSM) representation from a SystemC description.

The properties to be checked are specified in the standardized *Property Specification Language* (PSL). For the property check a variant of *Bounded Model Checking* (BMC) is used. Therefore, the property and the FSM representation are converted into an instance of *Boolean Satisfiability*

[1] A new version has been presented recently in [SKF$^+$09].

D. Große and R. Drechsler, *Quality-Driven SystemC Design*,
DOI 10.1007/978-90-481-3631-5_4, © Springer Science+Business Media B.V. 2010

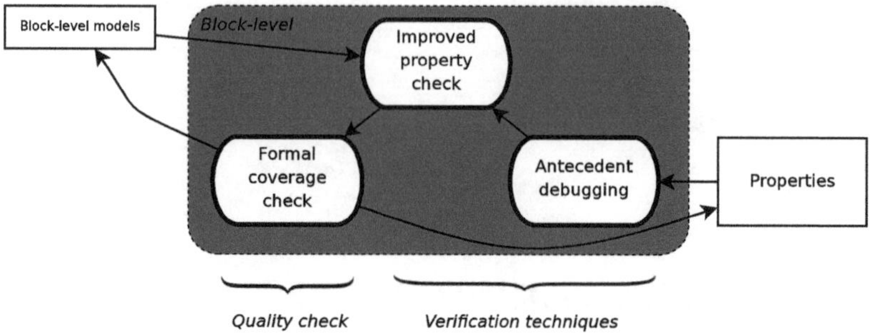

Figure 4.1. Block-level parts of enhanced SystemC design and verification flow

(SAT). Since the properties in this book are specified as implications and are formulated over a bounded time interval, they are proven if the corresponding SAT instance is unsatisfiable. Section 4.1 describes the BMC approach, the bounded PSL properties including their semantics and the implemented property checker in detail. Also several SystemC examples and their verification are discussed. The approach and achieved results have been published in [GD05b, DG05]. As property checking is often applied iteratively, a technique has been developed to accelerate the corresponding proofs. Here, iteratively means that the verification engineer usually specifies a property with a "strong" antecedent at the beginning of property checking. Then, this antecedent is stepwise weakened and the property check is performed again. For this scenario a speed-up of the underlying SAT proof is achieved. The technique is presented in Section 4.2 and has been published in [GD05a]. All mentioned approaches are summarized as the verification task *improved property check* in Figure 4.1.

Usually in the antecedent of a property the assumptions about the design environment are specified and joined by logical AND. However, if, e.g., the verification engineer focuses on a complex verification scenario in the antecedent of a property, the overall conjunction (possibly together with the design) may have no solution, i.e. the antecedent of the property is contradictory. Since in this case the property trivially holds, the consequent is never checked. Hence, this situation has to be avoided by identifying the reason. In Section 4.3 a fully automatic approach for debugging of a contradicting antecedent is presented. The method determines all minimal reasons for the antecedent contradiction and reveals if a conflict results from the property and/or parts of the design. The concepts of the debugging approach are similar to the approach presented Section 3.3. However the debugging problem that is considered here additionally takes contradictions with the design into account. The approach has been published in [GWKD09]. In Figure 4.1 the approach is denoted as *antecedent debugging*.

Finally, the verification quality at the block level is guaranteed by a formal coverage check presented in Section 4.4. The approach generates a coverage property for each considered signal. If the specified properties do not describe the signal's entire behavior, the coverage property fails and a counter-example is generated. A counter-example corresponds to an uncovered scenario, i.e. a verification gap. How to deal with verification gaps to achieve full coverage is discussed in detail. As final result the behavior of the design is determined under all circumstances, i.e. the property set completely specifies the behavior of the design. The approach has been published in [GKD07, KGD07, GKD08].

4.1 Property Checking

As described in the preliminaries *property checking* (or *model checking*) [CGP99] is a key verification technique to show whether a design satisfies the specification or not. In the last years especially *Bounded Model Checking* (BMC) [BCCZ99, BCC$^+$03] has become very successful in industrial practice [AKMM03, ADK$^+$05], since BMC overcomes the often observed memory explosion of classical model checking. BMC reduces the verification problem to a SAT problem and then searches for counter-examples in executions whose length is bounded by k time steps. If the resulting SAT instance is satisfiable a counter-example of length k has been found. However, BMC can only show that the design is free of errors for the given property up to the bound k. For proving a property, k has to finally reach the sequential diameter of the underlying FSM, which is infeasible for large circuits. Therefore, approaches for BMC and safety properties have been developed which can ensure completeness (see e.g. [SSS00, IPC03, WTSF04]). More precisely, the result of these approaches is a proven (or disproven property), in contrast to the "usual" result that no counter-example up to the bound k has been found.

The property checker which is presented in this section is based on a variant of BMC as proposed in [BS01, Joh02, BJW04, WTSF04]. It is characterized as follows: First, only properties over a fixed time interval – usually specified as implications – are allowed. Second, the restriction of the starting state for the unrolled circuit logic to the initial state is replaced by assumptions formulated explicitly in the antecedent of the property. As a result, the BMC problem consists only of a single SAT instance built by synthesizing the property and unrolling the design as many times as the property requires. If this SAT instance is unsatisfiable, the property holds.

Recently, in the context of SystemC property checking, some papers have been published. But still deriving a model for formal verification at the different abstraction levels is very difficult [Var07]. A first approach which uses a reachability analysis for checking LTL formulas of SystemC descriptions has been proposed in [DG02, GD03b]. However, this approach is limited to gate-level designs. A method that models a SystemC description as a labeled Kripke structure by partitioning the system into hardware and software parts has been presented in [KS05]. This technique allows proving properties using predicate abstraction [GS97] but does not take timing into account. In [HT05] an approach addressing model checking of TLM has been introduced. However, the design entry of this method is the *Unified Modeling Language* (UML) and only during the construction of the derived FSM some properties can be checked. Finally, the approach performs only simulation-based verification. Another approach has been proposed in [MMMC06]. This technique can extract structural and behavioral data from a SystemC design using the gcc front-end. After abstraction an intermediate model is built which can be used for model checking of generic properties. In [KEP06] a SystemC design is translated into Petri nets and CTL model checking is applied. However, the resulting Petri nets become very large even for small SystemC descriptions as the experiments show (several hours were necessary to verify simple properties for a small communication system).

This section is structured as follows. Section 4.1.1 reviews and formalizes important aspects of the used BMC variant. Afterwards in Section 4.1.2 the syntax and semantics of the PSL subset is described which is used here for specifying properties. Implementation aspects and a small example are discussed in Section 4.1.3. Then, Section 4.1.4 presents experimental results.

4.1.1 Bounded Model Checking

This section describes the principles of the used BMC variant. First, the properties over a finite time interval are formalized. Then, the BMC instance is defined.

For a considered sequential design let $(I, O, S, S_0, \delta, \lambda)$ be the corresponding Mealy machine. Furthermore, in the following $s^t \in S$ denotes the states at time point t, $i^t \in I$ the inputs and $o^t \in O$ the outputs at time point t, respectively. Then, a property is defined as:

DEFINITION 4.1 *A* property *over a finite time interval* $[0, c]$ *is a function*

$$p : (I \times O \times S)^{c+1} \rightarrow \mathbb{B}.$$

In practice, this function is specified as an implication, which is detailed in the next section.

For a sequence of inputs, outputs and states the value of

$$p(i^0, o^0, s^0, \ldots, i^c, o^c, s^c)$$

determines whether the property holds or fails on the sequence. Based on such a bounded property the corresponding BMC instance $b : I^{c+1} \times S \rightarrow \mathbb{B}$ is formulated. Thereby, the state variables of the underlying FSM are connected at the different time points, i.e. the current state variables are identified with

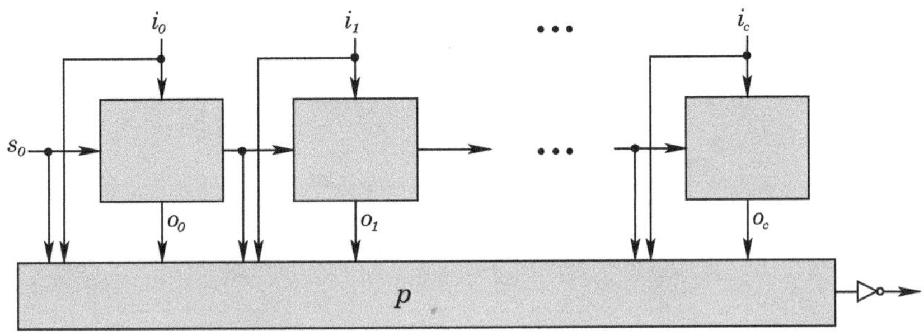

Figure 4.2. Unrolled circuit and property

the previous next state variables. This concept is called *unrolling*. In addition the outputs over the time interval are determined by the output function of the Mealy machine. Formally, we have:

DEFINITION 4.2 *For a Mealy machine* $M = (I, O, S, S_0, \delta, \lambda)$ *and a property* $p : (I \times O \times S)^{c+1} \rightarrow \mathbb{B}$ *over the time interval* $[0, c]$ *the corresponding* BMC instance *is given by:*

$$b(i^0, i^1, \ldots, i^c, s^0) = \bigwedge_{j=0}^{c-1} (s^{j+1} \equiv \delta(i^j, s^j)) \wedge \bigwedge_{j=0}^{c} (o^j \equiv \lambda(i^j, s^j)) \wedge \neg p$$

In Figure 4.2 the unrolled design and the property resulting in the defined BMC instance is depicted. To check the property the BMC instance is transformed into a SAT instance. As the property is negated in the formulation, a satisfying assignment corresponds to a case where the property fails, i.e. a counter-example has been found. In contrast to the original BMC as proposed in [BCCZ99] there is no restriction for the state s^0 in the first time frame during the proof. This may lead to *false negatives*, i.e. counter-examples that start from an unreachable state. In such a case these states are excluded by adding additional assumptions to the property. But, for the BMC variant as used here, it is not necessary to determine the diameter of the underlying sequential circuit. Thus, if the SAT instance is unsatisfiable, the property holds.

In the next section the PSL subset is described which is used in this book for specifying properties.

4.1.2 Property Language

Describing temporal properties can be done in many different ways, since there exist several languages and temporal logics (see also Section 2.3.2). We use the *Property Specification Language* (PSL) [Acc05]. PSL has been developed by Accellera, a consortium of several chip and EDA vendors. Meanwhile PSL has been standardized as IEEE standard 1850 [IEE05c].

```
property PROP =
  always(
    // antecedent
  ) -> (
    // consequent
  );
```

Figure 4.3. General structure of PSL property

For BMC we use only a subset of PSL. In this book each property is an implication of the form $always(A \rightarrow C)$. A is the *antecedent* and C is the *consequent* of the property. The motivation for the implication form is that the verification engineer defines assumptions about the design environment in the antecedent of the property and specifies the expected behavior under these assumptions in the consequent. In PSL this general structure is reflected as shown in Figure 4.3.

The antecedent and the consequent consist of expressions in temporal logic, i.e. propositional logic extended by temporal operators to reason about relations at different time points within the finite time interval $[0, c]$.[2] In PSL different flavors for the expressions exist, e.g. VHDL or Verilog. Since the designs are modeled in SystemC we use a SystemC flavor in the following. Thus, the propositional expressions and operators are taken from SystemC and C++, respectively. As temporal operators for a PSL property we have:

- `next[i](expr)`: The expression `expr` holds in `i` steps.
- `prev[i](expr)`: The expression `expr` holds `i` steps before the current time point.
- `next_e[a..b](expr)`: The expression `expr` holds at least once in the interval `[a,b]`.
- `next_a[a..b](expr)`: is the dual operator for the next_e operator, i.e. the expression `expr` holds always in the interval `[a,b]`.

In the following example a simple property is shown to illustrate some of the temporal operators.

EXAMPLE 4.3 *Figure 4.4 depicts the property* PROP. *This property is defined over the interval* $[0, 3]$. *The antecedent consists of three assumptions: The* `reset` *is disabled for three cycles (line 2),* `mode` *is* 2 *at time point* 0 *(line 3) and*

[2]Overall, the considered properties can be written in LTL as $G\varphi$ with $\varphi = A \rightarrow C$, where only the temporal operator X is allowed in the antecedent A and the consequent C.

```
1 property PROP = always(
2     next_a[0..2](reset == 0) &&
3     mode == 2 &&
4     high == 1
5   ) -> (
6     next[3](out) == 1
7   );
```

Figure 4.4. Example PSL property

`high` *is* 1 *also at time point* 0 *(line 4). Under these assumptions specified in the antecedent, the property requires in the consequent that* `out` *has to become* 1 *three cycles later (line 6).*

To define the semantics of PSL properties we recall the notion of paths in a Kripke structure informally (see also Definition 2.9). A path $\pi = \langle s^0, s^1, \dots \rangle$ in a Kripke structure corresponds to an execution of the Mealy machine. In the corresponding Kripke structure the labeling function $L(s^j)$ returns all input, output and state variables that are true at the particular time point j for a path. The semantics of the PSL properties is determined by the relation $\models_j$ which is defined in the following. For the propositional expressions we only give a minimal operator set. Further operators can be reduced to the given operators.[3]

DEFINITION 4.4 (SEMANTICS OF PSL PROPERTIES) *Let φ, ψ be formulas using only the propositional operators and the above mentioned temporal operators, $i, a, b \in \mathbb{N}$, and π be a path. Then the relation $\models_j$ is inductively defined as:*

$$
\begin{array}{lcl}
\pi \models_j \varphi & :\Leftrightarrow & \varphi \in L(\pi(j)) \\
\pi \models_j \neg\varphi & :\Leftrightarrow & \textit{not } \pi \models_j \varphi \\
\pi \models_j \varphi \wedge \psi & :\Leftrightarrow & \pi \models_j \varphi \textit{ and } \pi \models_j \psi \\
\pi \models_j \texttt{prev}[i](\varphi) & :\Leftrightarrow & \pi \models_{j-i} \varphi \textit{ for } j \geq i \textit{ and } i > 0 \\
\pi \models_j \texttt{next}[i](\varphi) & :\Leftrightarrow & \pi \models_{j+i} \varphi \textit{ and } i > 0 \\
\pi \models_j \texttt{next\_e}[a..b](\varphi) & :\Leftrightarrow & \exists k, a \leq k \leq b : \pi \models_{j+k} \varphi \\
\pi \models_j \texttt{next\_a}[a..b](\varphi) & :\Leftrightarrow & \forall k, a \leq k \leq b : \pi \models_{j+k} \varphi
\end{array}
$$

The described subset of PSL allows us to specify properties over a finite time interval. If the `prev[i]` operator is used in a property, one has to take care that the referenced time point does not become negative (this is the reason

[3]The same holds for arithmetic and relational operators, e.g. $+, -, <, \leq$, etc. These operators can be build by mapping them to respective circuit descriptions and synthesizing them down to the Boolean level.

for the condition $j \geq i$ in the definition above). During the construction of the property the verification tool checks whether the `prev[i]` operator is used incorrectly and terminates with an error in this case.[4]

Based on the semantic definition a PSL property of the form $always(\varphi)$ holds, if the formula φ holds for all executions of the FSM starting in an arbitrary state, i.e. in contrast to BMC as proposed in [BCCZ99] here also invalid executions – executions starting from an unreachable state – are considered. Formally, we have

DEFINITION 4.5 (VALIDITY) *Given a PSL property* $p = always(\varphi)$ *over the time interval* $[0, c]$*, we define*

$$\forall \pi \in \Pi : \pi \models_0 \varphi \Rightarrow \; p \; holds$$

If the antecedent of this implication is false, it has to be checked whether the starting state of the failing execution is reachable or not. In the former case a counter-example for the property has been found. In the latter case a false negative (see also discussion in Section 4.1.1) has occurred. However, the main strength of the used BMC variant is that properties can be proven without unrolling up to the diameter of the underlying FSM.

4.1.3 Implementation

In this section the property checker for SystemC is described. Before the details are given, the overall flow is outlined. After design implementation (coming from a refined system-level model) and formalization of the specification into temporal properties, the proposed approach works as follows (see also Figure 4.5):

1. The SystemC design is transformed into an internal FSM representation by the SystemC front-end.
2. A single property and the FSM representation is translated into a BMC problem.
3. The BMC problem is checked for satisfiability to decide whether the property holds or not.

These steps are now discussed in more detail.

FSM Representation

The property checker uses the SystemC front-end presented in [FGC+04]. For the generation of the FSM representation the last step of the overall procedure of the front-end has been extended (see also Figure 4.6). This step

[4] It is also possible to determine all time points in a pre-processing phase and then "moving" the whole property such that the first referenced time point becomes zero.

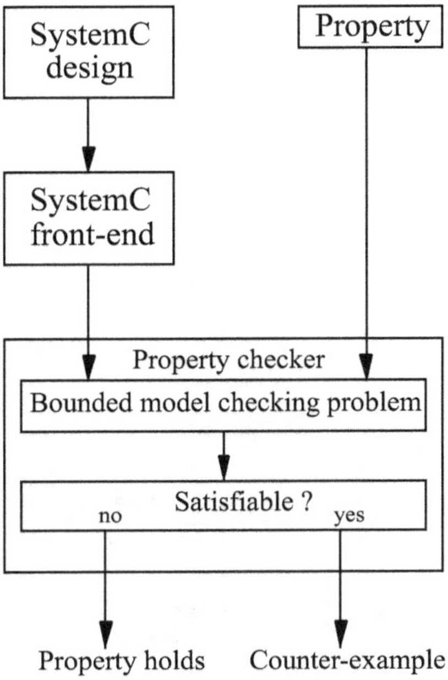

Figure 4.5. Property checking work flow

now generates additional information like for example data-types of variables, names of enum fields and data about hierarchy. This information is stored also in the resulting file.

BMC Problem and Property Checker

The property checker takes the FSM representation of the SystemC design and a property as input. Then the property is translated into an expression using only inputs, states, referenced internal signals and outputs of the SystemC design annotated with time points. The unrolled FSM representation and the property expression are converted into a bit-level representation. Here hashing and merging techniques for minimization are used, i.e. for example constants are automatically propagated and commutativity is exploited. The bit-level representation is given to the SAT solver zChaff [MMZ$^+$01] which has been integrated into the property checker. In case of a counter-example a waveform in VCD format [IEE01] is generated to allow for easy debugging.

To illustrate the transformation of a property and the unrolled FSM representation into a BMC instance we provide a small example. In Figure 4.7 a SystemC description of a 2-bit counter is shown. Besides the clock input the counter has a reset input `reset` and the output `out`. The current value of the counter is stored in `count_val`. Figure 4.8 depicts the underlying FSM of

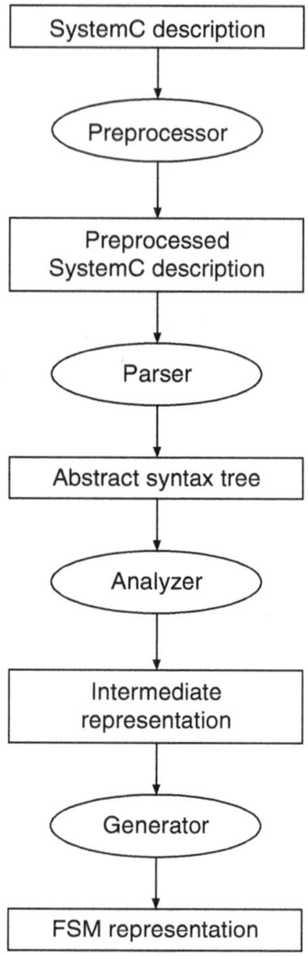

Figure 4.6. Transformation of a SystemC description into an FSM representation

the counter. Since the output and the state of the counter are identical only the states are shown in the figure. Basically this FSM is the result of the SystemC front-end. Besides the FSM other information of the SystemC design is stored, like e.g. data-types of variables. If the reset is represented by the Boolean variable r and `count_val` by the two state variables h (high) and l (low), the transition function of the FSM is given by:

$$\delta_h(r,h,l) = \neg r \wedge (h \oplus l)$$
$$\delta_l(r,h,l) = \neg r \wedge \neg l$$

For the counter three properties have been formulated. The first property `RESET` describes the reset behavior of the counter (see Figure 4.9). With the second

```
SC_MODULE( counter )
{
  sc_in_clk clock;
  sc_in<bool> reset;
  sc_out< sc_uint<2> > out;

  // counter value
  sc_uint<2> count_val;

  void do_count() {
    if (reset.read()) {
      count_val = 0;
    } else {
      count_val = count_val + 1;
    }
    out = count_val;
  }

  SC_CTOR( counter ) {
    SC_METHOD( do_count );
    sensitive << clock.pos();
  }
};
```

Figure 4.7. 2-bit counter

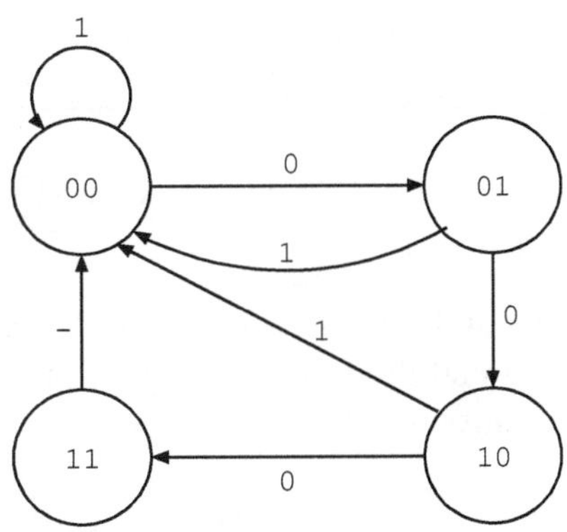

Figure 4.8. FSM of 2-bit counter

property `COUNT` the normal operation of the counter is characterized (see Figure 4.10). The last property states that the counter counts from three to zero (see Figure 4.11). Obviously these properties hold for the 2-bit counter. Now we consider the property `MODULO` in more detail. Since this property argues

```
property RESET = always(
    reset == 1
  ) -> (
    next[1](count_val) == 0
  );
```

Figure 4.9. Property RESET for the module `counter`

```
property COUNT = always(
    reset == 0 &&
    count_val < 3
  ) -> (
    next[1](count_val) == prev[1](
        count_val) + 1
  );
```

Figure 4.10. Property COUNT for the module `counter`

```
property MODULO = always(
    reset == 0 &&
    count_val == 3
  ) -> (
    next[1](count_val) == 0
  );
```

Figure 4.11. Property MODULO for the module `counter`

over the interval $[0, 1]$ the underlying FSM is unrolled only once ($c = 1$). For the state vector $s = (h, l)$ and the input r, the first part of the resulting BMC instance is (see also Definition 4.2):

$$\begin{aligned}\bigwedge_{j=0}^{1-1}(s^{j+1} \equiv \delta(i^j, s^j)) &= s^1 \equiv \delta(i^0, s^0) \\ &= (h^1, l^1) \equiv \delta((r^0), (h^0, l^0)) \\ &= (h^1 \equiv \neg r^0 \wedge (h^0 \oplus l^0)) \wedge (l^1 \equiv \neg r^0 \wedge \neg l^0)\end{aligned}$$

Based on this unrolling step the values of variables at time point 1 are available (e.g. `count_val` at time point 1 via h^1 and l^1). The part regarding the output is not shown because in the property the output of the counter is not used. The property MODULO corresponds to the following function:

$$p(\, i^0, o^0, s^0, i^1, o^1, s^1 \,) = \neg(\underbrace{\neg r^0}_{reset@0=0} \wedge \underbrace{h^0 \wedge l^0}_{count\_val@0=3}) \quad \text{(antecedent)}$$

$$\vee\, (\underbrace{\neg h^1 \wedge \neg l^1}_{count\_val@1=0}) \quad \text{(consequent)}$$

Finally, as explained above the BMC instance is converted into CNF and checked for satisfiability.

4.1.4 Experimental Results

The property checker has been implemented in C++. All experiments have been carried out on an Intel Pentium IV 3 GHz with 1 GB RAM running Linux. A run-time limit of 2 CPU hours has been set.

In a first example we studied a scalable hardware realization of the bubble sort algorithm. The SystemC description is shown in Figure 4.12.

This module implements the sort algorithm for eight data words. The bit size of each data word is determined by a `typedef`. Notice that the approach from [GD03b] did not support constructs, like e.g. typedefs or for-loops. In total, the correctness of sorting has been proven with nine properties. The first property `SORTED` ensures that the resulting sequence is ordered correctly, i.e. that the value of an output is greater or equal compared to values at outputs with smaller indices (see Figure 4.13).[5]

In Table 4.1 the results are given for the property `SORTED` and increasing bit sizes of data words (column *bit size*). The next two columns provide information about the SAT instance, i.e. the number of clauses and literals, respectively. In the last column the overall CPU time needed in CPU seconds is reported. Due to the heuristic nature of the SAT solver the proof time might vary slightly as can be seen in case of bit size 8 where the increase is not monotone. But in general the run-time needed increases with the bit size and is moderate even for larger bit sizes.

Finally, additional eight properties have been proven for the SystemC module `bubble`. These properties formalize that all input values of the module `bubble` can be found at the outputs. The summarized results for different bit sizes are shown in Table 4.2. Again the first column gives the bit size. In the next two columns details of a single SAT instance are shown. These numbers are identical for each of the eight properties, since the properties are symmetric, i.e. only the according input differs within the eight properties. The last

[5]For this property the antecedent is empty. However, a "1" is specified in the antecedent of the PSL property due to restrictions of our PSL parser.

```
typedef sc_uint<4> T;
SC_MODULE( bubble )
{
  sc_in< T >    in[8];
  sc_out< T >   out[8];
  T buf[8];
  void do_it() {
    for(int i = 0; i < 8; i++)
      buf[i] = in[i];
    for(int i = 0; i < 8-1; i++) {
      for(int j = 0;j < (8-i)-1; j++) {
        if( buf[j] > buf[j+1] ) {
          T tmp;
          tmp = buf[j];
          buf[j] = buf[j+1];
          buf[j+1] = tmp;
        }
      }
    }
    for(int i = 0; i < 8; i++)
      out[i] = buf[i];
  }
  SC_CTOR( bubble ) {
    SC_METHOD( do_it );
    sensitive << in[0] << in[1] << in[2] << in[3]
              << in[4] << in[5] << in[6] << in[7];
  }
};
```

Figure 4.12. Bubble sort

```
property SORTED = always(
    1
  ) -> (
    out[0] <= out[1] &&
    out[1] <= out[2] &&
    out[2] <= out[3] &&
    out[3] <= out[4] &&
    out[4] <= out[5] &&
    out[5] <= out[6] &&
    out[6] <= out[7]
  );
```

Figure 4.13. Property SORTED for module `bubble`

Table 4.1. Results for different input sizes of module `bubble` and property SORTED

Bit size	Clauses	Literals	CPU time (s)
4	6,390	14,458	17.18
8	12,754	28,894	286.93
16	25,482	57,766	125.25
32	50,938	115,510	560.48

Table 4.2. Results for different bit sizes of module `bubble` and input properties

Bit size	Clauses	Literals	CPU time (s)
4	6,298	14,262	58.49
8	12,570	28,502	681.52
16	25,114	56,982	845.76
32	50,202	113,942	3662.07

column provides the sum of the run-times for all eight properties. As can be seen, the correctness of the implementation of the bubble sort algorithm can be proven for up to 32 bits in 1 CPU hour.

While SystemC 1.x focused more on RTL descriptions, SystemC 2.x supports several constructs for system-level modeling. In this context channels are of high relevance. An important example of a channel provided with the SystemC distribution are FIFOs. In the refinement step, these FIFOs are then translated to the RTL. In a second series of experiments synchronous FIFOs with variable depth have been studied. The FIFO uses a register bank, a read pointer, a write pointer and a counter. It supports simultaneous read and write. Different properties have been developed which describe, e.g. the behavior after reset, no change of the FIFO content if no data is written to the FIFO and that the data is stored into the FIFO in case of a write access. For a bit size of 32 bits and increasing FIFO depths results are shown in Table 4.3. In the first and second column the depth of the FIFO and the property are given, respectively. In the next two columns details on the SAT instance are provided, i.e. the number of clauses and literals. Finally, the run-time is given in the last column. The results clearly show that also for high FIFO depths, i.e. FIFOs with more than 100 registers, the verification time needed is in the range of a few minutes. This demonstrates that even though it cannot be expected that complete systems can be checked, also complex system-level constructs can be formally verified using this approach.

Table 4.3. Results for different FIFO depths

Depths	Property	Clauses	Literals	CPU time (s)
8	reset	2,708	6,264	0.26
8	nochange	11,145	25,631	0.51
8	write	13,302	30,612	0.81
16	reset	5,181	12,025	0.52
16	nochange	22,309	51,327	1.78
16	write	26,158	60,248	2.75
32	reset	9,958	23,162	1.08
32	nochange	44,557	102,539	7.90
32	write	51,741	119,229	14.24
64	reset	19,343	45,051	2.35
64	nochange	88,865	204,531	40.64
64	write	102,680	236,670	58.63
128	reset	37,944	88,444	6.50
128	nochange	177,377	408,279	247.82
128	write	204,415	471,227	283.74

4.2 Acceleration of Iterative Property Checking

In SAT-based property checking the initial SAT instance is generated from the design description together with the property to be proven. Usually, the largest part will result from the unrolled design description. In comparison, the logic parts that come from the property are much smaller. From a practical perspective, during property checking as long as no design bug is found the design remains unchanged, but the verification engineer modifies and adds new properties. Thus, the property checker is used interactively. For the verification engineer on the one hand, proving becomes more easy if the assumptions in the antecedent of a property are very strong, i.e. the property is very restrictive and argues only over a small part of the design. On the other hand, such proofs are not very general. Hence in practice, the formulation of a property is an iterative process. For example, the verification engineer starts writing a property with strong assumptions. Then, the verification engineer stepwise weakens some of the assumptions to obtain a more general proof.

The basic idea is to exploit the iterative process of property checking. As described, only a very small part of the verification problem changes in consecutive property checking runs if the assumptions are weakened. Thus, recomputations can be avoided if learned information is reused for consecutive SAT problems. BMC as introduced in [BCCZ99] reduces the verification problem to a SAT problem and then searches for counter-examples in executions whose length is bounded by k time steps. For BMC, it has been suggested to reuse constraints on the search space deduced in instance k for solving the consecutive instance $k + 1$ faster [Sht01]. However, in [Sht01] this concept is only used during the proof of a single or several fixed properties.

In contrast to [Sht01], in the BMC variant that is used in the following two SAT instances for slightly different property checking problems are considered and information from the two properties with respect to the underlying design is utilized. This enables to reuse learned conflict clauses in the SAT instance of the consecutive property checking problem.

The approach has been integrated in the property checker presented in Section 4.1. Hence, all necessary information is available: For a design and a given property the property checker stores resulting conflict clauses and relevant information in a data base. In consecutive runs for a property and a derived version of the property (by weakening assumptions) some of the stored clauses can be reused. Typically, this makes the current instance easier to solve, since the search space is pruned by the reusable clauses. Besides this, recomputations of identical conflicts can be avoided. So a speed-up of the current proof can be expected. Reusable conflict clauses are such clauses that can be deduced by the intersection of the resulting clauses of the two property checking problems. We will show that these conflict clauses can be identified efficiently, if the information about the source of conflicts in terms of the property and a variable mapping is preserved. Experiments show that up to 100% of the clauses can be reused. This results in speed-ups of nearly a factor of 30 in our experiments.

The remaining part of the section is structured as follows: Section 4.2.1 provides the main flow for reusing conflict clauses in iterative PC. In Section 4.2.2 the formalization of the approach is presented. Experimental results demonstrating the benefits of the approach are described in Section 4.2.3.

4.2.1 Main Flow

In this section the main flow for reusing conflict clauses during iterative property checking is presented. In Figure 4.14 this flow is depicted.

At first the design and the property are compiled into an internal representation. In this step information to allow for a syntactic comparison between properties is stored in the data base (A). Then, the internal representation is converted into a BMC problem expressed as a CNF formula. While solving this SAT instance the references to the clauses that lead to a new conflict clause are stored in a data structure. After termination of the SAT solver this conflict clause information can be related to the expressions in the antecedent and the consequent of the checked property. Finally, this information is minimized and added to the data base (B). Now assume that property checking is repeated but the property has been weakened. This is detected (X) and before the BMC

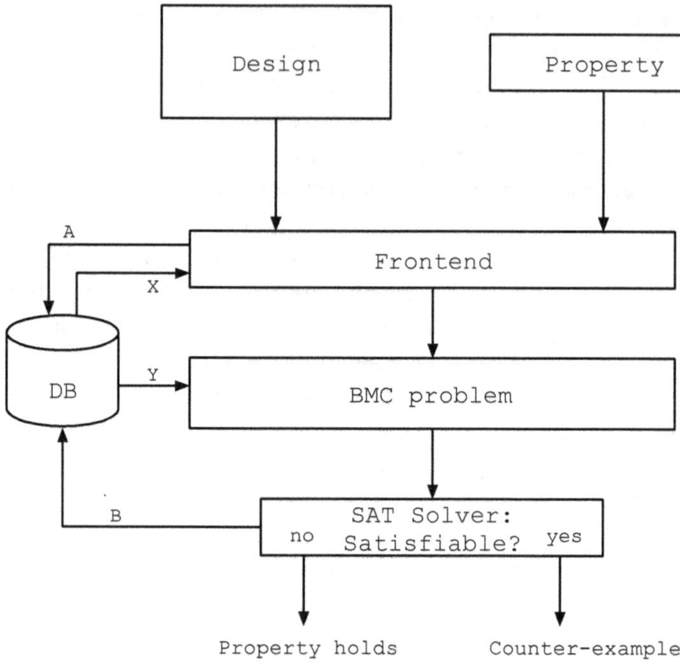

Figure 4.14. Property checking flow with reusing

problem is given to the SAT solver, conflict clauses are read from the data base. Then, they are analyzed and reused (Y), if possible. In the following, the approach is illustrated for an example.

EXAMPLE 4.6 *In Figure 4.15 the property* `lowestWins_2` *is shown. The property has been written for a scalable bus arbiter. The arbiter consists of* n *cells (here* $n = 3$*) and combines priority arbitration with a round robin scheme. The property* `lowestWins_2` *states that if exactly one token is set* (A_0) *and no cell is waiting (persistent signals)* (A_1) *and exactly the request* `req2` *is high* (A_2, A_3, A_4)*, then the corresponding acknowledge will be set in the same clock cycle.*

This property has been proven and the conflict clause information has been written in the data base as explained above. Now the question is raised whether all assumptions are necessary. A detailed analysis of the arbiter and the property shows that assumption A_4 *can be removed, since this assumption is too restrictive. (If* `req2` *is 1, the value of* `req3` *does not matter.) Proving this modified property is now faster because the re-computation of already learned conflicts is superfluous, since conflict clauses from the previous proof are reused.*

We continue with the formalization of the proposed approach.

```
property lowestWins_2 = always(
(A0)   (cell1.token + cell2.token + cell3.token) == 1 &&
(A1)   (cell1.pers + cell2.pers + cell3.pers) == 0  &&
(A2)   req1 == 0 &&
(A3)   req2 == 1 &&
(A4)   req3 == 0
) -> (
  ack2 == 1
);
```

Figure 4.15. Example property `lowestWins_2`

4.2.2 Reusing Learned Information

Let M be the set of clauses resulting from the translation of the design D, and let P be the set of clauses resulting from the property p. Then P can be partitioned into $P = A \cup C \cup R$, where A are the clauses from the antecedent, C from the consequent and R the clauses to "glue" the antecedent expressions and the consequent expressions of the property together. Now consider two consecutive runs of the property checker for the unchanged design D and for two properties p^F(first) and p^S(second). Assume that the property p^S has been derived from the property p^F by weakening some of the assumptions in the antecedent. Let $P^F = A^F \cup C^F \cup R^F$ be the resulting clauses of the property of the first run and $P^S = A^S \cup C^S \cup R^S$ the clauses for the second run, respectively. Further assume that the variables in P^S are renamed with a variable mapping function which maps a variable from the second set of variables V_S to the according variables of the variable set V_F from the first run. Then, the following holds:

1. $C^S = C^F$, since the consequents of properties p^S and p^F are equal.
2. $R^S = R^F$, since the variables to combine the antecedent and the consequent can be identified.
3. $A^S \subset A^F$, because the assumptions in the antecedent of p^S are weaker than the assumptions of p^F.

Since the clauses M of the design remain unchanged, only the clauses resulting from the two properties p^F and p^S have to be compared. We formulate the following theorem:

THEOREM 4.7 *For the corresponding set of clauses P^F and P^S of two consecutive property checking problems it holds:*

$$P^F \setminus P^S \quad = \quad A^F \setminus A^S$$

Proof

$$
\begin{aligned}
P^F \setminus P^S &= (A^F \cup C^F \cup R^F) \setminus (A^S \cup C^S \cup R^S) \\
&= (A^F \cup C^F \cup R^F) \setminus (A^S \cup C^F \cup R^F) && \text{(1. \& 2. from above)} \\
&= (A^F \cup C^F \cup R^F) \cap \overline{(A^S \cup C^F \cup R^F)} && \text{(set difference)} \\
&= (A^F \cup C^F \cup R^F) \cap \overline{A^S} \cap \overline{C^F} \cap \overline{R^F} && \text{(de Morgan)} \\
&= A^F \cap \overline{A^S} \cap \overline{C^F} \cap \overline{R^F} \cup && \text{(distributive law)} \\
&\quad\; C^F \cap \overline{A^S} \cap \overline{C^F} \cap \overline{R^F} \cup \\
&\quad\; R^F \cap \overline{A^S} \cap \overline{C^F} \cap \overline{R^F} \\
&= A^F \cap \overline{A^S} \cap \overline{C^F} \cap \overline{R^F} && \text{(empty intersections)} \\
&= A^F \cap \overline{A^S} && (A^F \cap C^F = A^F \cap R^F = \emptyset) \\
&= A^F \setminus A^S && \text{(set difference)}
\end{aligned}
$$

■

Based on Theorem 4.7 all conflict clauses can be reused which have *not* been learned from a conflict where clauses of $A^F \setminus A^S$ participated. In other words, we have to identify the conflict clauses which have been deduced exclusively from the intersection of the two consecutive property checking problems. This intersection is given by:

$$
\begin{aligned}
(M \cup P^F) \cap (M \cup P^S) &= M \cup (P^F \cap P^S) && \text{(distributive law)} \\
&= M \cup (A^F \cup C^F \cup R^F \cap A^S \cup C^S \cup R^S) && \text{(definition of } P\text{)} \\
&= M \cup (A^F \cup C^F \cup R^F \cap A^S \cup C^F \cup R^F) && \text{(1. \& 2. from above)} \\
&= M \cup ((A^F \cap A^S) \cup (C^F \cup R^F)) && \text{(distributive law)} \\
&= M \cup A^S \cup C^F \cup R^F && \text{(3. from above)}
\end{aligned}
$$

Thus, for each conflict clause of the first run the sequence of clauses which produced that conflict clause have to be determined. With this information we can exactly identify the source of the conflict in terms of the two properties p^F and p^S. This becomes possible, if we further know which clauses have been produced by the design, the individual expressions in the antecedent (separated at the logical ANDs) and the individual expressions of the consequent of both properties. Finally, for a conflict clause *cl* the minimal source information is stored which allows to check whether *cl* was produced by a clause of the design or by an antecedent or a consequent expression. Altogether it can be decided which conflict clauses of the first run can be reused to speed up the current proof.

4.2.3 Experimental Results

All experiments have been carried out on an AMD Athlon XP 2800+ with 1 GB main memory. The following experiments always consist of two steps. First, for a circuit a property with "overly" strong assumptions is proved. This is done with and without our approach to measure the time overhead. Next, we prove the same property but in a more general version, i.e. some of the assumptions in the antecedent of the property have been weakened. In this case we measure the speed-up that can be achieved by reusing conflict clauses.

In a first series of experiments we considered the scalable bus arbiter which has already been used in Example 4.6. The considered properties for the arbiter circuit are mutual exclusion of the outputs of the arbiter and the `lowestWins` property already described in Example 4.6. In Table 4.4 the overhead for our approach is given for different arbiter sizes (column *Cells*). In the second column the name of the considered property is shown. The next two columns provide information about the corresponding SAT instance. The run-time needed without and with our approach is given in column *std* and column *reuse*, respectively. The difference between the two given run-times is the time needed to store learned information into the data base. As can be seen the overhead is negligible, i.e. less than 1% of the run-time for the larger examples.

The achieved improvement of the proposed approach for the arbiter is shown in Table 4.5. In the weakened variant of the property `mutualexclusion` the assumption that no arbiter cell is waiting is no longer used. In case of the property `lowestWins_50` we follow exactly Example 4.6. The first seven columns give the same information as in Table 4.4. Because the considered properties have been weakened the resulting number of clauses and literals decreases. However, since for each property learned information can be found in the data base, conflict clauses can be reused. Thus, column *Reused Cl.* gives

Table 4.4. Overhead for arbiter

Cells	Property	Clauses	Literals	CPU time (s) Std	Reuse
100	mutualexclusion	240,776	541,742	9.15	9.57
100	lowestWins_50	161,399	363,193	14.15	14.49
200	mutualexclusion	961,576	2,163,542	176.65	177.78
200	lowestWins_50	642,799	1,446,393	588.30	590.45

Table 4.5. Acceleration for arbiter

Cells	Property	Clauses	Literals	CPU time (s) Std	Reuse	Reused cl. (%)	Speed-up
100	mutualexclusion	161,076	362,442	13.26	13.01	20.23	1.0
100	lowestWins_50	161,247	362,839	8.71	4.54	100.00	1.9
200	mutualexclusion	642,176	1,444,942	1078.80	343.77	6.23	3.1
200	lowestWins_50	642,347	1,445,339	656.35	22.70	100.00	28.9

Table 4.6. Overhead for FIFO

Size	Property	Clauses	Literals	CPU time (s) Std	Reuse
64	nochange	68,077	156,723	14.82	14.92
128	nochange	156,595	361,173	101.83	102.03

Table 4.7. Acceleration for FIFO

Size	Property	Clauses	Literals	CPU time (s) Std	Reuse	Reused cl. (%)	Speed-up
64	nochange	68,072	156,712	14.80	2.16	100.00	6.9
128	nochange	156,590	361,162	101.72	6.42	100.00	15.8

the percentage of reused clauses. In the last column the achieved speed-up is shown. As can be seen for the 100 cell arbiter in case of the property `mutualexclusion` no speed-up results. But for the three remaining examples a significant speed-up was obtained, i.e. up to nearly a factor of 30.

In a second series of experiments we studied FIFOs of different depths. In a property we prove that the content of a FIFO does not change under the assumption that no write operation is performed. In the initial version of this property it has also been assumed that no read operation is performed. Similar information as for the arbiter examples is provided in Tables 4.6 and 4.7, respectively. Also in this case for larger examples a speed-up of more than a factor of 10 can be observed.

Overall, the proposed approach significantly improves the run-time for proving properties in an iterative property checking flow.

4.3 Contradictory Antecedent Debugging for Property Checking

In practice, during the specification of a property the verification engineer formulates assumptions about the design environment in the antecedent of the property and joins them by logical AND. A typical example for such an assumption is to disable the reset for several cycles. However, for sophisticated properties formulated for complex designs the verification engineer may be confronted with the problem of a contradictory antecedent, i.e. the antecedent has no solution. Obviously such a situation has to be detected by the property checker, since otherwise the consequent would not be checked.

The underlying ideas of the debugging method presented in the following are similar to the approach introduced in Section 3.3.1. The difference is that in property checking a contradiction may also occur in combination with the design. This has to be taken into account by the approach and actually given as feedback to the user.

Typical scenarios that lead to a contradictory antecedent are

- Typos in an expression and/or temporal operator in the antecedent
- Misinterpretation of the specification or incorrect specification
- Bug(s) in the design, whereas the antecedent conforms to the specification
- Too strong assumptions about the design environment

The latter case can often be observed in practice, since the verification engineer intentionally specifies strong assumptions to understand a complex design. At the beginning it is much easier to focus on a certain design functionality instead of writing a general property.

A closer inspection of these scenarios reveals that two different kinds of a contradictory antecedent have to be distinguished:

1. The contradictory antecedent is solely caused by one or more conflicts of the antecedent expressions.
2. The contradictory antecedent results from one or more conflicts of the antecedent expressions *and* the design.

Related work to the described problem is discussed below. For BMC as used in this book, to test if the antecedent is contradictory is straightforward. Instead of checking the whole property, only the antecedent (for point 1) or the antecedent including the unrolled circuit logic (for point 2) is tested for satisfiability. However, in case of a negative answer, i.e. the SAT instance is unsatisfiable and hence we have a contradictory antecedent, the verification engineer has to identify what exactly causes the contradiction. As the debugging of a contradictory assumption is done manually, this might become a very time-consuming process.

In this section we present a fully automatic approach to analyze a contradictory antecedent. The proposed method can handle all scenarios as described above. Thereby, first the reasons of the contradictions in the property are given and in the second step the reasons with respect to the design. A reason is a conjunction of antecedent sub-expressions (and the design) that evaluates to 0. In addition, a reason is minimal in the sense that removing a sub-expression from the conjunction resolves the contradiction. Thereby, the method helps the verification engineer in debugging since he/she understands what causes the contradictions. The approach is based on a reformulation of the antecedent using new free variables such that sub-expressions of the antecedent can be disabled. From the assignments to the free variables the approach derives which sub-expressions are "non-relevant", i.e. they are never part of any contradiction. For the remaining sub-expressions the logical dependencies of the respective values of the free variables are analyzed and thereby minimal reasons for all contradictions in the antecedent are determined.

In this book temporal properties are specified as implications. If the antecedent of a property is contradictory such a property is said to be *vacuously satisfied* in the literature [BBDER97]. An *antecedent failure* – for the first time mentioned in [BB94] – has been considered as motivation to study the more general question: can a model or a property contain an error if model checking was executed successfully [BBDER97, BBDER01]. Searching for errors in this direction is called *vacuity detection* [BBDER97, KV99, PS02, AFF$^+$03, SDGC07]. Vacuity for temporal logic model checking is syntactically defined as follows: A formula φ is vacuously satisfied in a model M if φ is satisfied in M and there exists a sub-formula ψ of φ which can be replaced by any formula θ without changing the truth-value of φ in M. In [KV99] this definition was further extended to consider occurrences of sub-formulas. However, all these approaches only address the *detection* of vacuity which is accomplished here by checking if the antecedent (and the design) is unsatisfiable. In [CS07] the vacuity checks are further improved in two directions: First, it is proposed to perform some of the vacuity checks without the design. Second, redundancy in a property set is identified to tighten the specification. In [SDGC07] the authors speed up vacuity detection for BMC by using resolution proofs. As a result they are able to identify vacuous variables in properties. Again, both papers are not targeting the identification of reasons in case of a contradictory antecedent. The most related paper to the approach presented in the following is [BDFR07] where a method is proposed to identify a *Temporal Antecedent Failure* (TAF). The authors consider model checking of temporal implication properties specified as regular expressions. The proposed method computes a position in the regular expression that is a reason for a TAF. However, a position may involve a complex Boolean formula that cannot be further analyzed.

This section is structured as follows: In Section 4.3.1 the analysis approach for a contradictory antecedent is presented. After the definition of the problem, the reformulation of the antecedent based on a partitioning of the antecedent is introduced. Then, in Section 4.3.2 the analysis algorithm is described in detail. Section 4.3.3 provides the experimental results. Besides some smaller examples also a real life example in the context of a MIPS CPU is presented.

4.3.1 Analysis of Contradictory Antecedents

In this section we describe the method for determining the reasons of a contradictory antecedent. First, it is checked whether the antecedent is contradictory. In this case, the contradiction analysis is started. The approach is based on a partitioning of the antecedent expression into sub-expressions. A *reason* is a set of sub-expressions that is sufficient to cause a contradiction. With our algorithm, all reasons for a contradiction can be computed.

First, we give a definition of the problem. Then, we describe the reformulation of the antecedent expression that is performed in preparation for the algorithms discussed in the next section.

Contradictory Antecedent

DEFINITION 4.8 *For a given design with the transition relation T_δ*[6] *and a property $p = A \rightarrow C$ over the finite interval $[0, c]$ let $\bigwedge_{i=0}^{c-1} T_\delta(s_i, s_{i+1}) \wedge \neg(A \rightarrow C)$ be the corresponding BMC instance. Then, A is a* contradictory antecedent *of the property p, iff*

- *A always evaluates to 0 (considering all inputs, states and outputs used in p as free variables) or*
- *$\bigwedge_{i=0}^{c-1} T_\delta(s_i, s_{i+1}) \wedge A$ always evaluates to 0 (i.e. the contradiction(s) are caused by both, the antecedent A and the design)*

REMARK 4.1 *The fact whether the contradiction(s) are caused solely by the antecedent or by the antecedent and the design is important for two reasons: First, in the former case debugging is simpler as will be shown later. Second, the size of the SAT instance is much smaller if the design does not have be unrolled as in the second case.*

Reformulation of the Antecedent

We use a partitioning of the antecedent into several sub-expressions. This partitioning is motivated by the typical form of an antecedent, i.e. the antecedent consists of assumptions that are joined by logical AND. In addition, the chosen partitioning allows the identification of contradictions at low computational costs. A refinement can easily be done by performing our analysis again for the first result. The partitioning of the antecedent of a property is defined as follows.

DEFINITION 4.9 *Let $p = A \rightarrow C$ be a property with a contradictory antecedent A. Then, the antecedent A is partitioned into n sub-expressions $A_0, A_1, \ldots, A_{n-1}$ such that $A = A_0 \wedge A_1 \wedge \cdots \wedge A_{n-1}$, where the positions of the logical AND operators are derived from the antecedent according to the conjunction of different assumptions.*

[6]In the following the symbol T_δ always covers the transition relation as well as the output functions (see also Definition 4.2).

EXAMPLE 4.10 *Consider again the property* PROP *in Figure 4.4. The antecedent of this property is partitioned into the three sub-expressions:*

A_0 = '**next_a** [0..2](reset == 0)',
A_1 = 'mode == 2', *and*
A_2 = 'high == 1'.

Based on this partitioning, a *reason* in terms of our approach is a subset of all sub-expressions, that form a contradiction and hence have to be considered by the verification engineer for debugging. The definition for a reason is given as follows. In contrast to the contradiction analysis approach for constraint-based simulation (see Definition 3.10), here contradictions in combination with the design have to be taken also into account.

DEFINITION 4.11 *Let $p = A \rightarrow C$ be a property with the contradictory antecedent partitioned into $A = A_0 \wedge A_1 \wedge \cdots \wedge A_{n-1}$. Then a* reason *for the contradiction is a non-empty set $R \subseteq \{A_0, A_1, \ldots, A_{n-1}\}$ such that all sub-expressions $A_j \in R$ (either in combination with the design or not) form a contradiction, i.e.*

$$\bigwedge_{i=0}^{c-1} T_\delta(s_i, s_{i+1}) \bigwedge_{A_j \in R} A_j \qquad or \qquad \bigwedge_{A_j \in R} A_j$$

is not satisfiable, respectively. Additionally all reasons are defined to be minimal, *i.e. removing any sub-expression A_j from R resolves the contradiction.*

REMARK 4.2 *In some cases more than one reason for a contradictory antecedent can occur. If in this case only one conflict is fixed the antecedent is still contradictory. Thus, our approach computes* all *reasons and thereby allows the verification engineer to handle the debugging of the contradictory antecedent in one single step.*

To determine the reasons for a contradiction our algorithm uses a *reformulated antecedent* A'. This is done such that each sub-expression A_j can be disabled by the BMC tool and hence each contradiction can be resolved.

DEFINITION 4.12 *Let A be a contradictory antecedent. Then A is* reformulated *to A' such that*

1. *For each sub-expression A_j a new free variable e_j (called* enable variable*) is introduced and*

2. *A_j is substituted by the implication $e_j \rightarrow A_j$*

In this way $A = A_0 \wedge A_1 \wedge \cdots \wedge A_{n-1}$ is reformulated to $A' = (e_0 \rightarrow A_0) \wedge (e_1 \rightarrow A_1) \wedge \cdots \wedge (e_{n-1} \rightarrow A_{n-1})$. For the reformulated antecedent A' the following holds:

1. If e_j is set to 1, then the sub-expression A_j is enabled.
2. If e_j is set to 0, then the sub-expression A_j is disabled, because $0 \rightarrow A_i$ evaluates to 1 independently of A_j.

4.3.2 Algorithms and Implementation

Based on the introduced definitions and concepts, first the main algorithm for the antecedent debugging approach is presented. Then, the algorithm to compute the reasons is described.

Main Algorithm

All steps introduced above are summarized in Algorithm 4.1: First, it is distinguished whether A causes a contradiction or not (line 1). In the former case A is reformulated to A' according to Definition 4.12 (line 2). If a contradiction already occurs solely in A (line 3), at first the analysis is done without the design (line 4). For the further consideration all reasons that have been found here are excluded (line 5). After this, the analysis is performed with the design to determine the reasons for contradictions caused by both, the antecedent A and the design (line 6). Thus, in total all reasons are presented to the verification engineer. Since properties with contradictory antecedents hold independently of the consequent no explicit property checking is necessary. Hence, after analysis the algorithm terminates and *contradictory* is returned (line 7). In contrast, if no antecedent contradiction occurs (line 8) the property is checked (line 9).

Algorithm 4.1: doPropertyCheck(Transition relation T_δ, interval $[0, c]$, property p)

Result: $result \in \{holds, fails, contradictory\}$

1 **if** $(\bigwedge_{i=0}^{c-1} T_\delta(s_i, s_{i+1}) \wedge A \equiv 0)$ **then**
2   A' = partitionAndReformulate(A);
3   **if** $(A \equiv 0)$ **then**
4   contradictionAnalysis(A', 1, $[0, c]$);
5   excludeFoundReasons();
6   contradictionAnalysis(A', T_δ, $[0, c]$);
7   **return** *contradictory*;
8 **else**
9   **return** *checkProperty(p,* T_δ, $[0, c]$*)*;

Algorithm for Analysis of the Contradictory Antecedent

If the antecedent is contradictory, then our approach determines all reasons of the contradiction. This task is performed by Algorithm 4.2. In the following the single steps of this algorithm are described in detail.

The reformulation of the antecedent from A to A' as described in the previous section allows to enable/disable sub-expressions. The basic idea for the computation of all reasons is as follows. Since the enable variables are free variables, we can obtain an assignment to these variables such that the overall contradiction of the antecedent is resolved. Such an assignment contains information which assumptions cannot occur together. But from a single satisfying assignment we cannot conclude which expressions form a contradiction. This is illustrated in the following example:

EXAMPLE 4.13 *Consider the property* MYPROP *depicted in Figure 4.16(a). For simplicity assume that there are no contradictions of the antecedent in combination with the design. All assignments to the enable variables that resolve the contradiction are given in Figure 4.16(b). As can be seen from a single assignment it is not possible to conclude what is the reason of a contradiction.*

Therefore, all satisfying assignments for the enable variables are computed in Algorithm 4.2 (line 2). This is done using an *All Solution SAT* solver, i.e. once a solution has been found a *blocking clause* [McM02] is added to exclude the same solution for the enable variables from the remaining search space and the search for another solution continues. Each new solution of the enable variables is stored in a set $\mathcal{A}$ (line 3).

Using these solutions, for each sub-expression A_j it can be checked whether A_j is either self-contradictory or never part of a reason according to the following two observations[7]:

OBSERVATION 4.1 *If e_j is 0 for all solutions, then the respective sub-expression A_j is self-contradictory.*

OBSERVATION 4.2 *If the assignment of e_j is don't care for all solutions (i.e. the value of e_j can be either 0 or 1 in all solutions), then the respective sub-expression A_j is never part of a contradiction.*

Thus, each sub-expression A_j for which the respective enable variable e_j matches one of these two properties is either self-contradictory or never part of a contradiction.

[7]Both observations and the respective proofs can be directly transferred from the method presented in Section 3.3.1. Therefore, the proofs are not given here again.

```
property MYPROP =
  always(
    (A0) x == 1 &&
    (A1) x > 5  &&
    (A2) y == 0
  ) -> (
    next[1](o) == 1
  );
```

(a)

e_0	e_1	e_2
0	0	0
0	0	1
0	1	0
0	1	1
1	0	0
1	0	1

(b)

e_0	e_1	e_2
0	–	–
1	0	–

(c)

Figure 4.16. Simple example for contradiction analysis

Algorithm 4.2: contradictionAnalysis(Reformulated antecedent A', design T_δ, interval $[0, c]$)

Result: Set $\mathcal{R}$ of reasons

$\mathcal{A} = \emptyset$; `// Set of assignments for the` e_j `variables`

while *(find new assignment a for* e_j *in (*$\bigwedge_{i=0}^{c-1} T_\delta(s_i, s_{i+1}) \wedge A'$*))* **do**

  $\mathcal{A} = \mathcal{A} \cup \{a\}$;

$\mathcal{R} = \emptyset$; `// Set of reasons`

$\mathcal{E} = \emptyset$; `// Set of enable variables for detailed analysis`

n =getNumberOfEnableVariables(A');

for $(j = 0 \ldots n - 1)$ **do**

  if $(\forall a \in \mathcal{A} : a(e_j) = 0)$ **then**

    $\mathcal{R} = \mathcal{R} \cup \{\{e_j\}\}$;

  else if $(\forall a \in \mathcal{A} : a(e_j) = \mathit{don't\ care})$ **then**

    continue;

  else

    $\mathcal{E} = \mathcal{E} \cup \{e_j\}$;

foreach $(X \in \mathcal{P}(\mathcal{E}))$ **do** from smallest to the largest

  if $(X = \emptyset \ \mathit{or}\ |X| = 1)$ **then**

    continue;

  else if $(\exists X' \in \mathcal{R} : X' \subset X)$ **then**

    continue;

  else if $((\bigvee_{a \in \mathcal{A}} \wedge \bigwedge_{e_j \in X} e_j = 1) \equiv 0)$ **then**

    $\mathcal{R} = \mathcal{R} \cup \{X\}$;

printReasons($\mathcal{R}$);

EXAMPLE 4.14 *Consider again the property* MYPROP *shown in Figure 4.16(a). Figure 4.16(c) depicts a disjoint sum of the product cover derived from the BDD representation of the function from Figure 4.16(b). As can be seen the value of the enable variable* e_2 *is always don't care and hence we can conclude that this expression is never part of a contradiction.*

In both cases the respective sub-expressions do not have to be considered any longer. This early classification significantly reduces the number of subsets $R \subseteq \{A_0, A_1, \ldots, A_{n-1}\}$ to be checked as reasons.

In Algorithm 4.2 both observations are applied in line 8 and line 10, respectively. In the former case (e_j is 0 for all solutions and thus A_j is self-contradictory) the enable variable e_j is added as a single reason to a set $\mathcal{R}$ storing all reasons for the contradiction (line 9). Note that $\mathcal{R}$ stores the reasons in terms of the e_j variables, not in terms of the respective sub-expressions A_j themself. If the second observation holds (e_j is don't care for all solutions and thus A_j is never part of a contradiction), then this sub-expression can be skipped (line 11). For all remaining cases (line 12), e_j is stored in a set $\mathcal{E}$ including all sub-expressions (in terms of enable variables) which cannot be classified by the two observations and thus have to be considered in the detailed analysis (line 13).

The detailed analysis checks subsets consisting of the remaining enable variables (i.e. sub-expressions) for being a reason of the contradiction. The respective subsets X are obtained by creating the power set $\mathcal{P}(\mathcal{E})$ of $\mathcal{E}$ (line 14). Thereby, the empty subset $\emptyset \in \mathcal{P}(\mathcal{E})$ as well as all subsets $X \in \mathcal{P}(\mathcal{E})$ with $|X| = 1$ are omitted (line 15) since $\emptyset$ can never be a reason for a contradiction and all subsets containing only one element are already covered by the two observations. Furthermore, by ordering the subsets according to their cardinality, the smaller conjunctions of sub-expressions are checked first. In this way, by excluding all supersets of reasons determined so far (line 17), minimality is guaranteed.

For each remaining subset X (i.e. for each combination) the conjunction of the respective sub-expressions is tested for a contradiction. Therefore, all variables $e_j \in X$ are assigned to 1 to enable all respective sub-expressions of X. Then, the resulting cube is combined with a disjunction of all solutions $a \in \mathcal{A}$ as shown in line 19. If the cube of enable variables (i.e. the enabling of the respective sub-expressions) leads to a contradiction, then a reason has been found and thus, X is added to $\mathcal{R}$ (line 20). The final result of the algorithm is the set of all minimal reasons. From this set the algorithm outputs all sub-expressions of each reason since a direct link to the syntax tree of the PSL property is available (line 21).

In summary, the presented approach computes all minimal reasons $\mathcal{R}$ of a contradictory antecedent by exploiting

- The reformulation of the antecedent A to A' which allows the disabling/enabling of sub-expressions
- A two-stage problem formulation to distinguish between contradictions solely caused by the antecedent and contradictions of the antecedent in combination with the design
- The two observations which reduce the numbers of subsets to be considered, and
- A detailed analysis for the remaining enable variables which excludes supersets of already found reasons

In the next section the application of the algorithm to practical examples is studied.

4.3.3 Experimental Results

Motivating Example

As an example we will use the FSM depicted in Figure 4.17. There are four states encoded by two bits. The FSM falls back to state 00 when the signal `reset` is set to 1. Otherwise, it steps through the states in increasing order and wraps around to state 00 unless the signal `hold` is set.

EXAMPLE 4.15 *Consider the property shown in Figure 4.18. In the antecedent it is assumed that there is a reset at time point* 0 *(line 2), the state should be* 10 *at the next time point (line 3) and there is no reset and no hold during the cycles 0 to 4 (lines 4 and 5, respectively), which leads to a contradiction. For the analysis, the antecedent is split into the four expressions* $A_0 \ldots A_3$

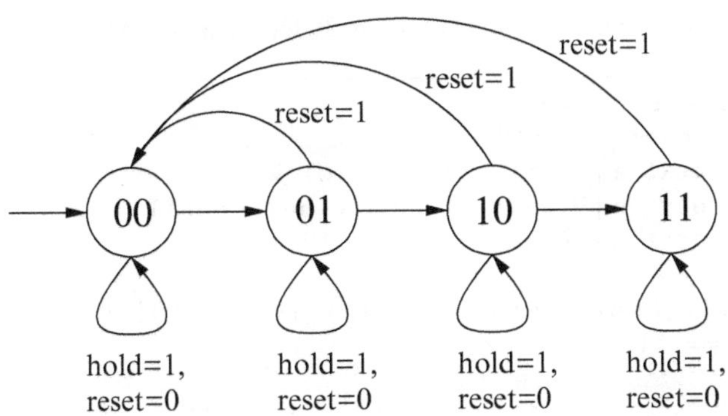

Figure 4.17. FSM

mentioned above. It is reported that A_3 *is irrelevant and that there are two different reasons for the contradiction. The first reason is* $R_1 = \{A_0, A_2\}$. *This is because the expression* $A_2 = \texttt{next\_a}[0..4](reset == 0)$ *implies that there is no reset at time point 0 which contradicts expression* A_0. *The second reason is* $R_2 = \{A_0, A_1\}$, *because after the reset at time point* 0 – *as demanded by* A_0 – *the state will be* 00 *at time point* 1, *which contradicts* A_1.

In case of the `next_a` *statements the analysis can still be refined, as the expression* $\texttt{next\_a}[i..j](x)$ *can be rewritten as a conjunction* $\texttt{next}[i](x)$ && $\texttt{next}[i+1](x)$ && ... && $\texttt{next}[j](x)$. *By splitting up the* `next_a` *operator the analysis points exactly to the expression related to time point 0 for the first reason.*

We consider another more complex example in the following.

EXAMPLE 4.16 *Consider the property in Figure 4.19. The antecedent is a conjunction of four expressions* $A_0 \ldots A_3$. *It is assumed that there is a reset at time point 0 followed by 5 cycles with no reset (lines 2 and 3). Furthermore there may not be two consecutive cycles with the hold signal activated (line 4). Finally, at time point 5 it is assumed that the FSM is in state* 01 *(line 5). When checking this property it is reported that its antecedent is contradictory. The single reason that is given by our approach is the conjunction*

```
property P1 = always(
(A0)   reset == 1 &&
(A1)   next( curr_state == "10" ) &&
(A2)   next_a[0..4]( reset == 0 ) &&
(A3)   next_a[0..4]( hold == 0 )
) -> (
   next[5]( curr_state == "00" )
);
```

Figure 4.18. PSL property for Example 4.15

```
1 property P2 = always(
2 (A0)   reset == 1 &&
3 (A1)   next_a[1..5](reset == 0) &&
4 (A2)   next_a[0..5](!(hold && next(hold))) &&
5 (A3)   next[5](curr_state == "01")
6 ) -> (
7   next[6](curr_state == "01") || next[6](curr_state
      == "10")
8 );
```

Figure 4.19. PSL property for Example 4.16

of all four expressions and the design. This means that removing any of the assumptions would remove the contradiction. In this case it is the combination of the antecedent with the functionality of the design that makes the scenario impossible.

RISC CPU

We applied our approach during the formal verification of a RISC CPU. The CPU implements parts of the MIPS instruction set architecture [PH04]. It is based on a five-stage pipeline and contains 32 general purpose registers. The overall design has a gate count of approximately 300,000.

During the design process BMC has been applied for early debugging of the basic functionality of the CPU. For this purpose relatively restrictive properties are written to check aspects of the design that have recently been implemented. One of these properties is shown in Figure 4.20. The property checks that the load word (LW) instruction works properly for a certain case. The antecedent is divided into seven expressions A_0 to A_6 (lines 14–21). It states that there is first a NOOP (no operation) instruction (A_0) to avoid control hazards

```
1  assign LW_op   = "100011";
2  assign NOOP_op = "00000000000000000000000000000000";
3  assign Instr1  = idata;
4  assign Instr2  = next(idata);
5
6  assign Op = Instr2.range(31, 26);
7  assign Rs = Instr2.range(25, 21);
8  assign Rt = Instr2.range(20, 16);
9
10 assign Addr = next[4]( mem.memaddr );
11 assign Data = next[4]( mem.rmemdata );
12
13 property LOAD = always(
14 (A0)  Instr1 == NOOP_op &&  // not in shadow of branch
15 (A1)  Op == LW_op &&        // execute load instruction
16 (A2)  Rt  != "00000" &&     // do not write to const. $0
17 (A3)  Data == 127 &&        // read value 127
18 (A4)  Addr == 13 &&         // from RAM address 13
19
20 (A5)  next_a[0..6]( !reset ) && // no reset
21 (A6)  next_a[0..6]( !ex )       // no exception raised
22 ) -> (
23    next[6]( reg.reg[Rt] ) == Data
24 );
```

Figure 4.20. Property for load instruction

(e.g. a taken branch). Then, the NOOP is followed by a load instruction (A_1), the target register is not the constant register (A_2) and the value 127 should be read from memory address 13 (A_3 and A_4). Furthermore it is assumed that during the execution of the instructions there is no reset and no exception raised (A_5 and A_6). Under these assumptions it should be proven that the value is saved into the correct register when the LW instruction finishes (line 23).

During BMC, a contradictory antecedent was reported and the analysis approach was started. It revealed a single reason corresponding to the subexpressions A_1, A_4, A_5, A_6. Furthermore, the contradiction is not caused solely by the antecedent, but in combination with the design. With this result, the problem could be figured out quickly. The design always raises an exception when there is an unaligned load access, i.e. when the address is not a multiple of 4. So the scenario described in the property – loading from address 13 without raising an exception – is impossible.

The experiments have been run on an Intel Xeon CPU with 3 GHz and 32 GB main memory under the Linux operating system. The time spent for the detection of the contradictory antecedent was 6.24 CPU seconds. The detailed analysis took another 50.86 CPU seconds, involving 121 SAT solver calls to collect the assignments to the enable variables. Note that the design of 300,000 gates has to be unrolled 6 times due to the time interval $[0, 6]$ of the property. This result clearly shows that the antecedent debugging can be performed very fast.

In summary, the debugging problem of a contradictory antecedent has been considered. This situation occurs if the verification engineer focuses on sophisticated scenarios where a manual design and property review process was necessary before. The introduced approach allows an automatic debugging of a contradictory antecedent and identifies whether a contradiction results solely from the antecedent and/or parts of the design. Hence, the debugging time is reduced significantly.

4.4 Analyzing Functional Coverage in Property Checking

Today, formal verification is standard in many industrial design flows. One prominent technique in this context is *Model Checking* (MC) [CGP99]. As already explained in Section 4.1 due to significant improvements in the tools for SAT solving, SAT-based MC methods like BMC [BCCZ99] can be applied to large designs and are widely used in industry [AKMM03, WTSF04, ADK+05, BBM+07].

In MC the functional properties are specified in temporal logic. Thus, each property describes parts of the circuit's behavior unambiguously. However, the most important question that arises here is: "Have I written enough properties?" [KG99]. In simulation-based verification computing coverage is well understood since coverage corresponds to an activation of signals or source code during the execution of input vectors. Also formal techniques have been used to analyze testbenches with respect to functional coverage [FD04]. For MC estimating coverage is more difficult since the correspondence of coverage is not obvious because the complete underlying *Finite State Machine* (FSM) is traversed during the proof. Nevertheless there exist several notions of coverage for model checking [CKV03]. For CTL-based approaches the covered states of the FSM are computed [HKHZ99, JPS03]. By *mutation-based coverage* modifications are applied to the FSM and it is checked whether this is detected by the specified properties [CKV01, CKV03]. But most of the existing metrics suffer from complexity problems as in the case of CTL model checking or are still very hard to use by verification engineers in practice.

An approach for measuring the coverage for BMC has been proposed in [FD06]. There the focus is mainly to relate errors in the design to the source code level. A component is considered covered, if there is at least one property that is invalidated, if this component is changed. The notion of "change" is described using a multiplexor construct based on the ideas of [AVS$^+$04]. The approach cannot specify exact functional coverage. Instead, components are identified that influence the behavior of the circuit. Thus, the approach cannot present the uncovered scenarios in form of counter-examples.

In [Cla06, Cla07] a method for coverage analysis of safety property lists has been presented. This method works only on the specified LTL properties without using the design. The properties are synthesized into a checker circuit that has exactly one output. This output is 1 iff all properties hold. Then, the checker circuit is duplicated. The two checker circuits use the same variables except for the output signal that is analyzed in the current step. Based on this construction the method can identify a forgotten case, i.e. a trace where a particular output signal is not constrained by the properties. Therefore the method looks for traces where both checker outputs are 1, but the values of the analyzed output signal differ at exactly one time point. The presented method is efficient because the design is not needed. However, if the method has identified a forgotten case there is no information about the circuit behavior, as the design is not regarded in the check. Furthermore, the method needs to introduce a new LTL construct for specifying that a signal is allowed to be unconstrained in a certain case, while our method is built on top of the standard PSL language.

In [BBM$^+$07] a technique to find gaps in a verification plan has been proposed (for more details see [Bor09]). It determines whether every possible input scenario corresponding to a sequence of operations of the DUV can be covered by a chain of properties that predicts the value of states and outputs at every point in time. The properties are linked up in a property graph which is used to carry out the completeness analysis. The properties which have to be specified for this approach are so-called operation properties. Such an operation property captures a single design transaction which is basically a transition between high-level design states. However, the notion of operation properties is very specific to the commercial approach behind [BBM$^+$07]. Instead, our method concentrates on revealing single concrete traces for simple designs. Furthermore, the techniques presented here can be applied to any set of properties without the need to specify a property graph.

In this section we propose a practical approach for the analysis of coverage in BMC as introduced in Section 4.1. The basic idea is the following: First, for each output o of the design all proven properties are identified which involve o. Then, it is checked whether there exists a scenario where o is *not* determined by the set of properties. Here "not determined" means that an input and state assignment has been found, where no consequent of the set of properties specifies the value of o unambiguously. We show that this idea can be integrated easily in a BMC verification tool. Furthermore, the approach automatically generates uncovered scenarios in form of counter-examples. Analyzing these counter-examples and adding corresponding properties allows the verification engineer to stepwise close the coverage gap.

As a non-trivial example we study the formal verification and coverage analysis of a RISC CPU. At first the block-level verification is considered. The verification and the coverage tests are described in detail by means of examples. Several cases that we observed to be the reason of a coverage gap are identified and it is explained how these gaps can be closed. During this process also the run-times for verification and coverage analysis are studied. Furthermore, we investigate the analysis of coverage on a higher level. Based on the results of the complete block-level verification we consider the RISC CPU at the top level.[8] Typically, at this level properties for each CPU instruction are formulated. In such a property the exact behavior of all involved hardware blocks with respect to the considered CPU instruction is specified. In other words the effect that results from the execution of an instruction including the communication of hardware blocks is checked. We show that the coverage approach can be used to guarantee coverage at this level. On the basis of a detailed example the suggested notion of higher level coverage based on proven correct instructions is discussed and the costs for the coverage check are analyzed, too. Following a certain property style is helpful for achieving full functional coverage by reducing the number of uncovered scenarios.

The remaining part of this section is structured as follows. Section 4.4.1 describes the general idea of the coverage approach. In Section 4.4.2 the details on the coverage property are provided. Then, in a case study in Section 4.4.3

[8]This is possible since the considered RISC CPU has a moderate gate count and hence can be handled formally also at the top level.

we demonstrate the coverage approach for the formal verification of a RISC CPU. In Section 4.4.4 we discuss how the verification flow can be improved by the presented techniques.

4.4.1 Idea

After proving a set of properties the verification engineer wants to know if the properties describe the complete functional behavior of the circuit. Thus, typically the properties are manually reviewed and the verification engineer checks that properties have been specified for each output (and important internal signals) which prove the expected behavior in all possible scenarios. Here, the goal of our approach is to automatically detect scenarios – assignments to inputs and states – where none of the properties specify the value of the considered output.

This idea is realized by the generation of a coverage property for each considered output. If this coverage property holds, there does not exist a scenario where the value of the output o is not determined by the properties. It is shown that the union of all properties that involve the output o admit no behavior else than the one defined by the circuit. This is done by introducing a multiplexor for each bit that is driven by the output o and the inverted value of o. Then the coverage check is performed by proving that the multiplexor is forced to select the original value of o, assuming all involved properties. In the following section this is described in more detail.

4.4.2 Coverage Property

To analyze the coverage we generate a coverage property for each output o of the design. This coverage property for the output o is constructed as follows:

1. The set of properties P_o that involve the output o is identified.

2. The maximum time point t_{max} is determined. The time point t_{max} is defined as the latest time point in the consequent of all properties in P_o, at which o is constrained.

3. A multiplexor is inserted for each bit of the considered output o in the original design. Let o consist of the single bit signals $o_0, \ldots, o_{n-1}$. Then, the bit o_i of the output is connected to the data input d_1 of the i-th multiplexor, whereas the negation of o_i is connected to the data input d_0. If o is a single bit signal, then we add the new input sel to the design that is connected to sel_0 (which is the select input of the multiplexor). Otherwise we add $sel_0, \ldots, sel_{n-1}$ as new inputs and the signal $sel = \bigwedge_{i=0}^{n-1} sel_i$.

4. The output o is renamed to o_orig and the name of the output of the inserted multiplexor is set to o. Thus, in the following all properties that

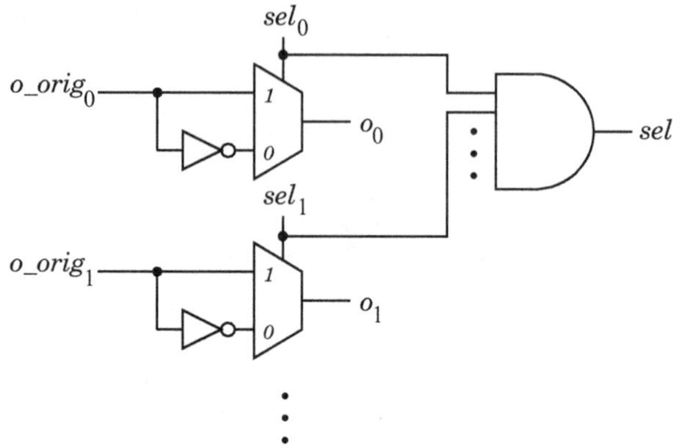

Figure 4.21. Insertion of the multiplexor

use o are dealing with the output of the multiplexor instead of the originally considered output. The overall transformation for the single bits of o including the multiplexor insertion from step 3 is depicted in Figure 4.21.

5. Now the coverage property for the considered output o is generated. In the antecedent, all properties of P_o are assumed. Possibly a property p_i has to be shifted by applying the next operator such that output o in the consequent of p_i is constrained at the maximum time point t_{max}. If the output o is constrained in the consequent of the property p_i at m time points with $m > 1$, then p_i is splitted into m separate properties, i.e. each such property consists of the antecedent of p_i and m different consequents. For each property p_i the resulting properties are summarized as $\hat{p}_i$. Furthermore, in the antecedent of the coverage property the signal sel is set to 1 during the time interval $[0, t_{max} - 1]$. This guarantees that in all properties $\hat{p}_i$ the original output o is used for all time points up to $t_{max} - 1$. In the consequent of the coverage property we force the signal sel to be 1 at time point t_{max}. More formally the coverage property is

$$\left(\bigwedge_{i=1}^{|P_o|} \hat{p}_i \wedge \bigwedge_{t=0}^{t_{max}-1} X_t sel = 1 \right) \rightarrow X_{t_{max}} sel = 1,$$

where X_j denotes the application of the next operator for j times.

Following these steps we have formulated the coverage analysis problem as a BMC problem. In the following theorem the soundness of the approach is considered.

THEOREM 4.17 *If the coverage property for the considered output o holds, then o is covered by the properties (given as the set P_o).*

Proof: We show this by contraposition: if the output o is not covered, then the coverage property fails. In the following we denote the output of the multiplexers with o_M. All properties $\hat{p}_i$ hold due to construction. If the output o is not covered, then the value of o_M at time point t_{max} is not uniquely determined by the properties $\hat{p}_i$. In contrast, if the value of o_M is determined by the properties $\hat{p}_i$, then for each input and state assignment there is a property that predicts the value of o_M. Hence, the value of o_M and o_orig have to be identical since the predicting property has been proven on the design. Thus, in this case sel is 1. Now, if the properties do not predict the value, then there exists at least one assignment o'_M for output o_M, that differs from the original value o_orig in at least one bit. But this is equivalent to the fact that the select signal for this bit can be set to 0 at time point t_{max}, thus selecting the value o'_M without invalidating any of the properties $\hat{p}_i$. As a consequence, there exists a counter-example with $sel = \bigwedge_{i=0}^{n-1} sel_i = 0$ at time point t_{max}, which means that the coverage property fails. ■

Complete coverage in terms of our approach is achieved by considering all outputs of a circuit. If all outputs are successfully proven to be covered by the properties, then the functional behavior of the circuit is fully specified.

Examples

As a first example consider the 1-bit memory shown in Figure 4.22. If the signal *we* (write enable) is set to 1, the flip-flop is updated with the value of the input *din*. Otherwise it keeps its value.

To verify an implementation of this simple memory cell, a PSL property has been specified (see Figure 4.23). It states that whenever *we* is set, the value of *din* can be seen at the output *dout* one cycle later. The property holds.

We can now check whether the behavior of the memory cell is covered by this property. Therefore, a coverage property is generated (see Figure 4.24). In line 2 a special command enclosed in a comment instructs the verification tool to insert a multiplexor at signal *dout* with a select signal named *sel*. In the

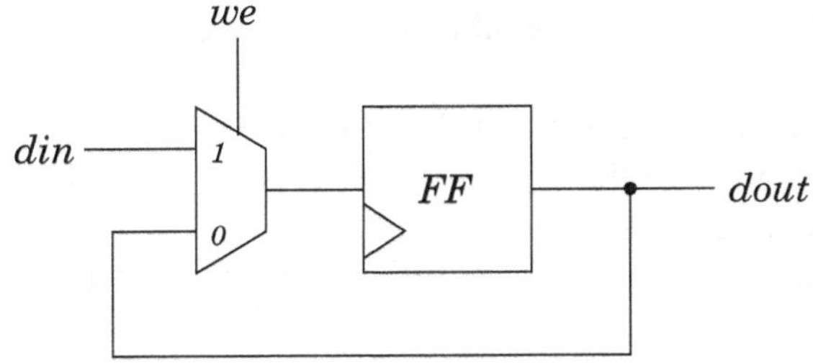

Figure 4.22. 1-bit memory

```
property WRITE =
  always(
    we == 1
  ) -> (
    next(dout) == din
  );
```

Figure 4.23. PSL property for the 1-bit memory

```
1 property COV =
2   // @insertMuxForSignal: dout sel
3   always(
4     ((we == 1) ? (next(dout) == din ) : 1) &&
5     sel == 1
6   ) -> (
7     next(sel == 1)
8   );
```

Figure 4.24. Coverage property for the 1-bit memory

antecedent of the coverage property it is assumed that property WRITE holds (line 4).[9] Furthermore, it is assumed that the *sel* signal is set to 1 in the first cycle. Thereby the original value of *dout* is routed to the output. Under these assumptions it has to hold that the select signal is 1 in the next cycle (line 7).

As a result, the coverage property fails and a counter-example is generated which is shown in Figure 4.25. The case that has not been covered can be deduced from the trace. Apparently it has not been specified how the memory cell behaves if the signal *we* is set to 0. As a consequence the *sel* signal can be set to 0 in the second cycle without violating the property WRITE. After adding an appropriate property like the one in Figure 4.26, the output is fully covered. The additional property NO_CHANGE states that the output remains unchanged as long as the write enable signal is set to 0.

As a second example we consider a FIFO of depth 3 that filters some value by setting the output to 0 if the last three inputs have been 1. See Figure 4.27 for the implementation of the FIFO. The two basic properties for the FIFO are shown in Figure 4.28. In the first property the regular shifting of the FIFO is proven if the content of the FIFO is different from three times 1. The second

[9]This is expressed in PSL using the C-like ?-operator for an if-then-else construct. The property $A \rightarrow C$ is transformed to $A \ ? \ C : 1$ due to syntactical restrictions of our PSL parser.

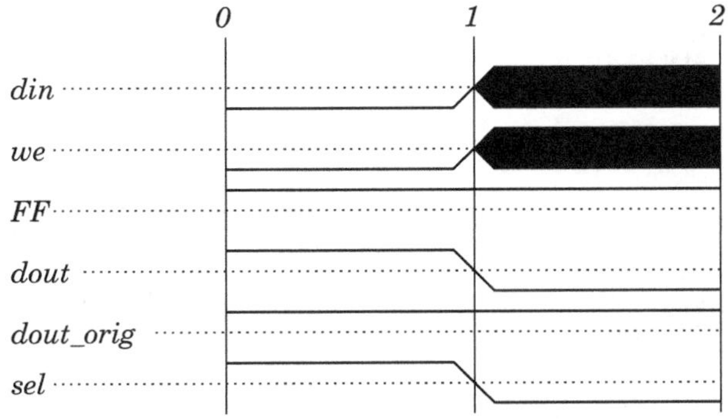

Figure 4.25. Counter-example for coverage of the memory cell

```
property NO_CHANGE =
  always(
    we == 0
  ) -> (
    next(dout) == dout
  );
```

Figure 4.26. Additional property for the 1-bit memory

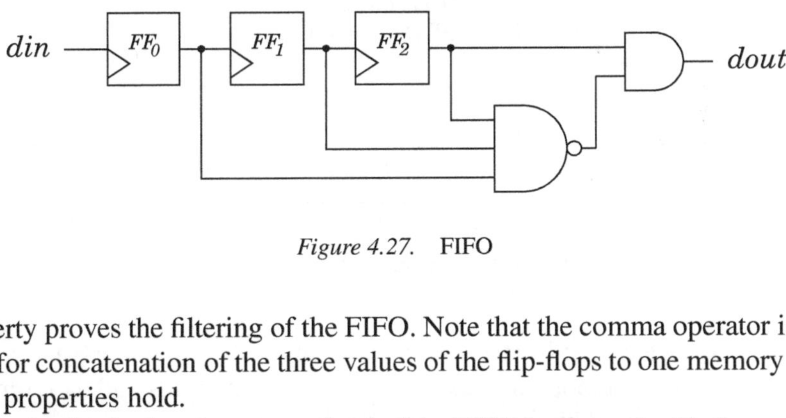

Figure 4.27. FIFO

property proves the filtering of the FIFO. Note that the comma operator is used here for concatenation of the three values of the flip-flops to one memory word. Both properties hold.

We check whether the output *dout* of the FIFO is covered with the coverage property given in Figure 4.29. Again the multiplexor is inserted with the special command in line 2. For the FIFO and the given properties t_{max} is 3.

Thus, the property PROPAGATE is assumed without shifting because o is already constrained at time point t_{max}. However, the property FILT has to be shifted such that the output *dout* is constrained in this property at time point t_{max} (see line 9). The last expression in the antecedent forces the select input to be 1 up to time point 2 (the **next_a** operator constrains the following expression to hold at every time point during the specified interval; see line 10). In the

```
property PROPAGATE =
always(
  next[3]( (FF0, FF1, FF2) != "111" )
) -> (
  next[3]( dout ) == din
);

property FILT =
always(
  (FF0, FF1, FF2) == "111"
) -> (
  dout == 0
);
```

Figure 4.28. PSL properties for the FIFO

```
property COV =
// @insertMuxForSignal: dout sel
always(
  // PROPAGATE
  (next[3]( (FF0, FF1, FF2) != "111" )?
  (next[3]( dout ) == din) : 1)&&

  // FILT property was shifted
  next[3]( (((FF0, FF1, FF2) == "111") ? (dout == 0) :
        1) ) &&
  next_a[0..2]( sel == 1 )
) -> (
  next[3]( sel == 1 )
);
```

Figure 4.29. Coverage property for the FIFO output

consequent we want to show that the select input is 1 at time point $t_{max} = 3$. The verification tool reports that the coverage property holds and therefore the FIFO output is covered by the two properties given.

Based on these two examples the generation of the coverage property for simple and more complex properties with respect to several time points has been shown. In the following our approach is studied on a RISC CPU.

4.4.3 Experimental Results

In this section we provide experimental results for the complete verification and coverage analysis of a RISC CPU, moving from the level of single hardware blocks to the instruction set on the top level. The distinction between these levels of abstraction is not a consequence of the coverage approach, but comes from the verification of hierarchical designs. In this methodology single hardware blocks, that might have been reused from former designs, are verified first before reasoning about the behavior of the complete circuit. All experiments that are reported have been carried out on an AMD Athlon XP 2800+ with 1 GB main memory under the Linux operating system. In the following, we first give some basic data of the CPU.

RISC CPU

In Figure 4.30 the main components of the RISC CPU are shown.

The CPU is designed as a Harvard architecture. The data width of the program memory and the data memory is 16 bit. The size of the program memory is 4 kByte and the size of the data memory is 128 kByte. The length of an instruction is 16 bit. We only briefly describe the five different classes of instructions in the following:

- 6 load/store instructions (movement of data between register bank and data memory or I/O device, loading of a constant into high- or low-byte of register)
- 8 arithmetic instructions (addition/subtraction with and without carry, left/right rotation and shift)
- 8 logic instructions (bit by bit negation, bit by bit exor, conjunction/disjunction of two operands, masking, inverting, clearing and setting of single bits of an operand)

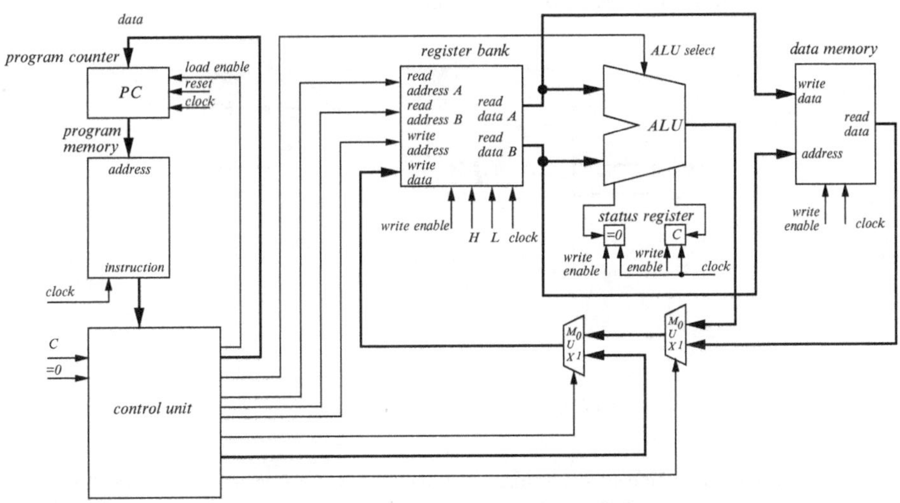

Figure 4.30. Structure of the RISC CPU including data and instruction memory

- 5 jump instructions (unconditional jump, conditional jump, jump on set/cleared carry or zero flag)
- 5 other instructions (stack instructions push and pop, program halt, subroutine call, return from subroutine)

The overall design has a size of approximately 22,000 gates and includes about 1,700 state bits.[10]

Block-Level Verification

In order to guarantee the correct behavior of the RISC CPU, it is verified using BMC. In a first step, for each of the hardware blocks it is checked, whether the input/output behavior of the implemented circuit matches the specification. Therefore a number of properties has been formulated in PSL.

As the *program counter* (PC) will serve as an example, it is described first. Figure 4.31 shows the PC with all its inputs and outputs.

The PC has an internal 11 bit register `pc` which holds the current program address. The address is shown at the output `pcout`, while the output `pcinc` shows the current address increased by 1. The PC is reset to address 0 by setting the input `reset` to 1. If the load enable input `le` is set to 1, the PC is loaded with the address from input `din`. Otherwise it is increased by 1 in every cycle if the PC is enabled, which means that the enable input `en` is set to 1.

In Figure 4.32 some of the properties for the PC can be seen. The first property `RESET` checks the correct behavior after a reset. The second property `INC` checks that the PC is increased if it is enabled, there is no reset, there is no load, and if the end of the address space has not been reached yet. The third property `LOAD` checks the load functionality of the PC.

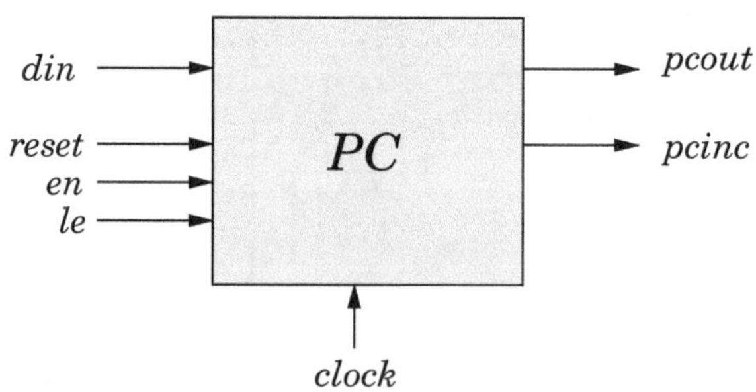

Figure 4.31. Program counter

[10]In the verification model the data and instruction memories have been reduced to 64 and 32 words, respectively. This is done for complexity reasons since otherwise the corresponding SAT instances become too large if the complete memories are included. Technically, the descriptions of both memories can be easily scaled down by resizing the respective array sizes and ignoring the higher bits of the addresses.

```
property RESET =
  always(
    reset == 1
  ) -> (
    next( pcout == 0 && pcinc == 1 )
  );

property INC =
  always(
    reset == 0 && le == 0 &&
    pc < 2047
  ) -> (
      next( (prev(en) == 1) ?
        (pcout == prev(pc) + 1) :
        (pcout == prev(pc))
      )
  );

property LOAD =
  always(
    reset == 0 &&
    le == 1
  ) -> (
      next(
        (prev(en) == 1) ?
        (pcout == prev(din)) :
        (pcout == prev(pc))
      )
  );
```

Figure 4.32. Properties for the program counter

For all hardware blocks of the CPU properties have been specified in a similar way. In Table 4.8 the results of the block-level verification are shown. The first column gives the name of the hardware block. The second column provides the number of properties that have been written for the respective block. In the last two columns the total CPU time for the verification and the maximal used memory during the verification are given. As can be seen, the verification can be carried out very fast using BMC.

Block-Level Coverage

Coverage Analysis. If all properties hold, the coverage check can be performed in a next step. Following the approach described in Section 4.4.2, for each single output of each hardware module it is checked, whether its behavior is specified unambiguously by the properties. Therefore a coverage property is generated for each output.

Table 4.8. Costs of block-level verification

Block	#p	CPU time (s)	Max. mem (MB)
ALU	18	4.30	15
Program memory	2	1.21	21
Data memory	2	4.52	41
Register bank	5	1.22	15
Program counter	4	0.10	8
Stack pointer	6	0.09	8
Control unit	19	0.23	8

```
1  property PCOUT_COV =
2  // @insertMuxForSignal: pcout select
3  always(
4    // RESET
5    ((reset == 1) ?
6     (next(
7        pcout == 0 && pcinc == 1
8      )) : 1) &&
9
10   // INC
11   (((reset == 0) && (le == 0) &&
12     (pc < 2047)) ?
13    (next(
14       (prev(en) == 1) ?
15       (pcout == prev(pc) + 1) :
16       (pcout == prev(pc))
17     )) : 1) &&
18
19   // LOAD
20   (((reset == 0) && (le == 1)) ?
21    (next(
22       (prev(en) == 1) ?
23       (pcout == prev(din)) :
24       (pcout == prev(pc))
25     )) : 1) &&
26
27   select == 1
28 ) -> (
29   next(select == 1) // covered?
30 );
```

Figure 4.33. Coverage property for the program counter

Figure 4.33 shows the coverage property for the output pcout of the PC. In line 2 the multiplexor needed for the coverage check is inserted using the already described special command. The original output pcout is replaced by

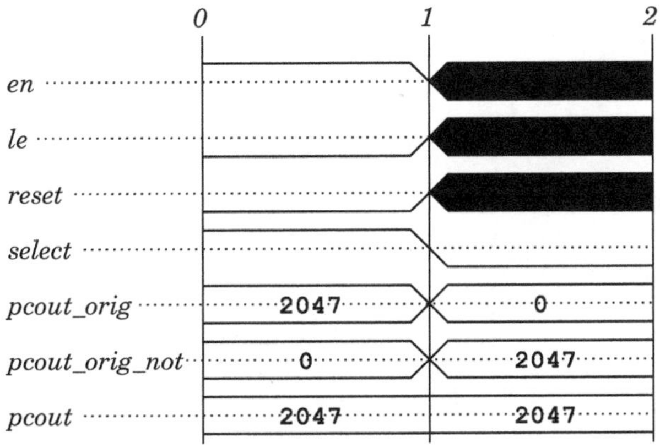

Figure 4.34. Counter-example for program counter coverage

the multiplexor construct – consisting of a multiplexor for each output bit – and is renamed to `pcout_orig`. Now the original value `pcout_orig` is routed to the output `pcout` iff the signal `select` is set to 1. Thus, all properties that use `pcout` are now treating the output of the multiplexers instead of the originally considered output (see also Figure 4.21). In lines 4 to 25 the original properties are assumed. Furthermore, it is assumed that at time point 0 the select signal is set to 1 (line 27), and thus we are dealing with the original value of the circuit on output `pcout` at time point 0.

Under these assumptions we want to prove that the select signal has to be 1 at time point 1 as well (line 29), meaning that the output is determined in any case. It appears that the coverage property fails. Figure 4.34 shows a counter-example that has been generated by the verification tool. From the trace it can be concluded which scenario has not been specified by the properties discussed in the previous section. As can be seen in the figure, none of the properties covers the case that the PC is enabled, there is no reset or load, and the PC points to the end of the address space.

At the same time, the trace gives information on the actual behavior of the circuit in the unregarded case. Signal `pcout_orig` gives the original value of `pcout`. Obviously, the PC starts over at address 0 when it exceeds the highest possible address.

Before closing this gap we present the results of the coverage analysis phase in Table 4.9. In the same way every hardware block of the CPU is checked based on the coverage approach. The first column of Table 4.9 gives the name of the module. In the second column the number of generated coverage properties is provided. Column *cov* reports whether all outputs were covered. If not, the last column provides the solution. As can be seen we found three gaps in total. The details on closing these gaps and gaps in general are discussed next.

Table 4.9. Results of coverage analysis

Module	# p	Cov	Solution
ALU	17	No	Added property
Program memory	2	Yes	–
Data memory	2	Yes	–
Register bank	4	Yes	–
Program counter	3	No	Excluded states
Stack pointer	3	No	Excluded states
Control unit	19	No	Added 2 properties, excluded states

Closing the Gap. If a gap is found using the presented coverage approach, there are different ways how to deal with it. It is possible that the verification engineer has in fact forgotten to check a certain scenario. In this case the properties have to be completed until coverage is achieved. For the RISC CPU we found that the properties for the ALU did not specify the value of the carry bit in case of a logical operation. Therefore, we added an according property. In the same way we had to add two properties in order to cover the behavior of the control unit in terms of our approach (see Table 4.9). For example it was not properly specified how the I/O interface behaves during a reset.

It is also possible that some scenarios have been left out intentionally, possibly because the specification itself is incomplete. In this case the assumptions of the coverage property can be extended to exclude these states explicitly. Referring to the block-level coverage analysis example, the specification did not define the behavior of the PC at the end of the address space. It is left to the programmer of the CPU to avoid an address overflow. This is expressed by excluding the state 2047 in the coverage property. Thereby, the program counter was fully covered. In Table 4.9 this procedure is denoted as "excluded states". As can be seen in the table, three harmless gaps have been found. For example one of the gaps in the control unit was related to inactive parts of the data path. In these cases the coverage was completed by excluding the respective states directly in the coverage properties.

In total by the presented coverage approach we found three coverage gaps. Following the described steps we achieved full coverage on the block-level.

Computation Costs. To compare the effort for verification and coverage the results of the final full coverage proof for each block are shown in Table 4.10. The first column gives the name of the hardware block. In the second column the number of outputs are given that have been checked for coverage in the

Table 4.10. Costs of block-level coverage

Block	#o	CPU time (s)	Mem (MB)
ALU	3	10.72	30
Program memory	1	1.33	23
Data memory	1	3.52	38
Register bank	2	0.48	17
Program counter	2	0.05	9
Stack pointer	1	0.04	8
Control unit	24	0.06	9

Table 4.11. Costs of instruction set verification

Category	#p	CPU time (s)	Max. mem (MB)
Load/store	6	86.98	94
Arithmetic	8	676.44	144
Logical	8	51.25	79
Jump	5	13.79	69
Other, reset	6	22.95	80

respective block. The last two columns show the run-time and the memory needed for the coverage check. As can be seen the run-times and memory requirements for coverage check and verification (see above) are in the same order of magnitude.

Top-Level Verification

Based on the successful verification of all involved hardware blocks, the instruction set of the RISC CPU is formally verified. A property has been formulated for each of the 32 instructions that checks if the effects of the instruction meet the specification. Typically, these properties affect all of the hardware blocks.

We only summarize the results of the verification, see Table 4.11. A detailed presentation for the ADD instruction can be found in the context of the HW/SW co-verification approach in the next chapter in Example 5.4 (Section 5.2.3). The first column gives the category of the verified instructions. The number of properties for the respective category can be found in the second column. The last two columns give the total run-time and the maximum memory needed during verification. Note that for the considered design the time interval of most of the properties is $[0, 1]$. Thus, the model has to be unrolled for two time steps.

Top-Level Coverage

Coverage Analysis. In contrast to block-level verification the properties for the instructions of the RISC CPU do not consider single outputs or signals. In fact the instruction set verification involves different hardware blocks and their communication. Therefore, the notion of coverage at this level is not as clear as for single hardware blocks. Obviously it is not sufficient to prove the coverage of the outputs of the CPU because the input/output interface is only affected by few instructions. To be sure that the properties form a complete specification of the circuit's behavior, the state holding elements have to be considered as well. If all state bits of the circuit are uniquely determined at any point in time, its behavior is fully covered in terms of our approach. We justify this approach by the fact that the circuit is equivalent to an FSM and by covering all state bits we describe the transition function in a unique way. Note that for the RISC CPU most of the outputs of a block become a next state input of a state element at the top-level.

As an example the status bits of the RISC CPU are considered – the zero flag and the carry flag indicate the result of the last logical or arithmetic operation, respectively (see also Figure 4.30). Among the properties, for top-level verification there are 32 properties for the instructions and a reset property. In order to achieve full coverage with this partitioning of the properties, *every* property has to define the value of the status bits, regardless whether the respective instruction changes the flags or not.

As an example consider the property for the jump instruction in Figure 4.35. It states that whenever there is no reset (line 6) and the current instruction is a `JMP` (line 7), then the program counter is set to the target address in the next cycle (line 11). This obviously describes the correct behavior of the jump instruction. However, in order to achieve full coverage the property must also specify what the jump instruction does *not* do. This is expressed in lines 14 to 21 which state that in the next cycle the status bits and the content of the registers remain unchanged.

As a consequence, all properties have to be assumed in the coverage property for the status bits. In this way, a large monolithic property is generated including *all* instructions. It could be proved that under all circumstances the value of the status bits is unambiguously defined by the instruction set properties.

Table 4.12 gives the results of the top-level coverage analysis. The first column shows the state bits and signals that have been tested for coverage. The second column reports whether the initial coverage check has been successful or not. In the latter case the solution to reveal the coverage gap is shown in the last column. As can be seen, we found some gaps in the coverage of the remaining state bits and the global outputs of the I/O interface. The steps taken to close these gaps are discussed in the next section.

```
assign OPCODE = instr.range(15,11);
assign DEST   = instr.range(10,0);

property JMP =
always(
  reset == 0 &&
  OPCODE == "11110"
) -> (

  // jump to target address
  next( pc.pc == DEST ) &&

  // no side effects
  next(
    (stat.C == prev(stat.C)) &&
    (stat.Z == prev(stat.Z)) &&
    (reg.reg[0] == prev(reg.reg[0])) &&
    (reg.reg[1] == prev(reg.reg[1])) &&
    [...]
    (reg.reg[7] == prev(reg.reg[7]))
  )
);
```

Figure 4.35. Property for the jump instruction

Table 4.12. Results of top-level coverage

Signal	Cov	Solution
Carry bit	Yes	–
Zero bit	Yes	–
Stack pointer	No	Adopted properties, excluded states
Program counter	No	Adopted properties
Constant register	Yes	–
General purpose register	No	Excluded states
I/O signals	No	Added property

Closing the Gap and Property Style. As for the block-level coverage discussed before there are different solutions for the gaps on top level. To achieve full coverage for the state holding elements of the program counter and the stack pointer, most of the instruction set properties had to be adopted to the style already mentioned in the previous section. For all instructions except jump and conditional branches we had to specify that the program counter increases by one. Similarly, we specified that all instructions except push, pop and subroutine calls do not change the stack pointer.

Table 4.13. Costs of top-level coverage

Signal	CPU time (s)	Mem (MB)
Carry bit	54.04	105
Zero bit	38.58	96
Stack pointer	4.40	88
program counter	4.42	88
Constant register	4.16	87
General purpose register	1,771.16	166
I/O signals	4.62	88

For the coverage of the stack pointer we excluded the states that correspond to a stack overflow or underflow, respectively. This has been done because the specification did not define the behavior for these cases, so it is left to the programmer to avoid these situations.

As the I/O signals are only affected by two instructions, we decided to define a new property that states that the I/O interface is in an idle state unless one of these two instructions is performed. After these changes we achieved full coverage on top level in terms of our approach.

Computation Costs. The costs for the final top-level coverage analysis are shown in Table 4.13. For most of the signals the coverage check could be carried out very fast. Most effort is spent for the status bits and the general purpose registers. However, the coverage proofs for the carry bit and the zero bit were still faster than some particular instruction set properties. Only the check for the registers required a significantly higher run-time. We assume that this is due to the fact that the instructions perform many different logical and arithmetic operations on the registers which have to be taken in account during the coverage check.

In summary, we have shown that the complexity to prove functional coverage is manageable with our approach. This is possible by breaking down the overall problem into several block-level and top-level proofs. Thereby each single proof in turn is carried out using BMC.

4.4.4 Discussion

In this section we discuss the contributions of the proposed approach. We show how the formal verification flow is improved using our method. The enhanced formal verification flow is illustrated in Figure 4.36. Note that the new parts of the flow are indicated in dashed lines. Starting from the specification the design is implemented. In parallel properties are formulated to prove that the design meets the specification using BMC. Based on the verification

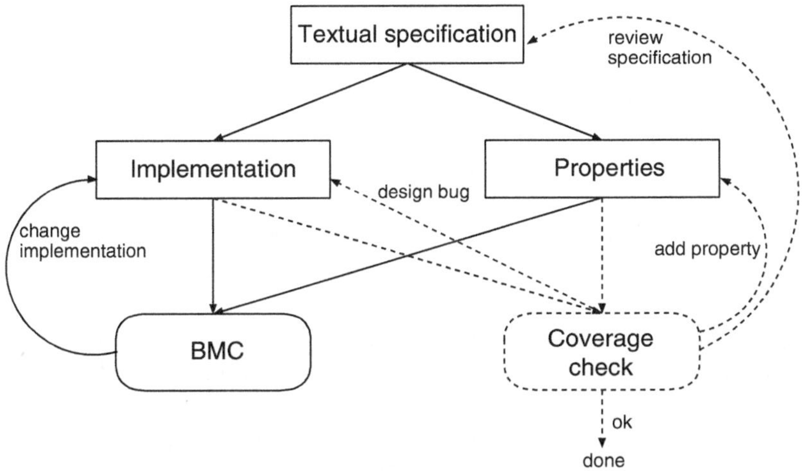

Figure 4.36. Enhanced verification flow

results the implementation is changed until the properties hold. In contrast to the traditional formal verification flow now a coverage check can be performed fully automated. If all properties cover the entire behavior of the design in terms of the coverage approach, then the properties form a complete and unambiguous specification of the design. This complete verification can be achieved stepwise by closing all coverage gaps that are being revealed by the approach.

Whenever a coverage check fails and a counter-example is generated for the uncovered scenario, the verification engineer has to decide whether the gap has to be considered harmless. The counter-example provides information on the actual behavior of the circuit in the uncovered scenario. If the behavior does not meet the specification, then there is a bug in the design. If the behavior conforms to the specification, then there is a coverage gap and the verification has to be completed by adding a property.

But it is also possible that the specification itself is incomplete and the verification engineer has left out certain scenarios intentionally. Another reason might be that the verification engineer only wants to check that a certain subset of all possible scenarios is covered by the properties. In these cases the unspecified scenarios can be excluded from the coverage check by an additional assumption in the coverage property. After handling of the gap the algorithm has to be rerun.

In any case the approach provides valuable feedback for the verification engineer and enables him/her to reason about the uncovered scenarios. In this way it supports design understanding and helps to improve the quality of the verification.

4.5 Summary and Future Work

For block-level verification improved property checking, automatic antecedent debugging and formal coverage analysis were considered in detail in this chapter.

At first, a property checking approach based on a variant of BMC has been introduced. To prove a property for a SystemC description the description is unrolled and together with the property converted into a SAT instance. The advantages of the approach have been shown in experiments. Since property checking is applied in an iterative way, i.e. often the assumptions in a property are stepwise weakened, a technique to accelerate the corresponding SAT proofs has been developed. For different examples a significant speed-up has been documented.

Next, an automatic debugging approach for a contradictory antecedent in property checking has been proposed. This situation occurs if the verification engineer checks sophisticated scenarios. The introduced approach distinguishes between a contradiction solely caused by the antecedent and contradictions that result from the antecedent and the design. As a result all minimal reasons are computed. Hence, the manual debugging process is replaced by an automatic method which reduces the debugging time significantly.

Finally, the presented block-level verification techniques are complemented by an approach to guarantee the verification quality. The proposed method analyzes functional coverage in property checking. This allows to automatically identify uncovered scenarios (so-called coverage gaps), i.e. scenarios where non of the specified properties define the behavior of the considered output unambiguously. After closing all of the gaps the highest possible verification quality is obtained since the behavior of the design is determined under all circumstances.

Directions for future work include the derivation of word-level information to further speed up the proofs. First general steps have been made already in this direction (see e.g. [WFG+07, SKW+07]) but the specific properties of SystemC have to be investigated. Furthermore, for the acceleration approach of iterative property checking as well as for the antecedent debugging approach UNSAT cores can be exploited. Another important research direction is property checking at the top level and system level, respectively. While proving properties at the top level needs better scalability of the proof techniques, property checking at the system level requires deriving suitable formal models as well as devising new high level verification methods. First steps to specify properties at the system level without relating expressions at clock cycles have been published recently [TVKS08]. Besides this, also methods to analyze coverage for such approaches have to be investigated.

Altogether the described verification techniques and the complementing formal coverage check allows to deliver a high quality system at the block level.

Particularly, in the proposed quality-driven design and verification flow the achieved result of 100% correct blocks builds the foundation for the following verification tasks that are considered in the next chapter: techniques are presented to bridge the gap between the block level and the system level, i.e. the communication of the completely verified blocks is the focus of top-level verification.

Chapter 5

TOP-LEVEL VERIFICATION

After complete formal verification at the block level – based on the techniques presented in the previous chapter – this chapter addresses the verification at the top level. The top-level verification task is required since large systems cannot be handled completely by formal methods due to complexity reasons. Thus, as introduced by the proposed design and verification flow, block-level verification is carried out first and then top-level verification starts on top of the high quality result, i.e. 100% correct proven blocks. Hence, the techniques that are presented in the following focus on the verification of the communication between the proven blocks.

The parts of the proposed design and verification flow that are covered here are depicted in Figure 5.1. In black the top-level parts are shown whereas in light gray the dependencies to the system level and the block level are illustrated, respectively. The dependencies are explained below when the respective top-level task is described.

First, in Section 5.1 an approach for checker generation is introduced. The basic idea is to embed properties directly into the top-level SystemC description after appropriate transformations have been carried out. Then, the embedded properties are checked during simulation at the top level. Especially properties for communication between different blocks are verified using this approach. In Figure 5.1 this verification task is denoted as *checker generation*. As can be seen for ensuring the verification quality the coverage check at the system level is used. Besides simulation, the checkers are also synthesizable and can be utilized for on-line tests after fabrication. The approach has been published in [GD04, DG05].

In addition to checker generation also formal verification at the top level is investigated for a special class of systems. For embedded systems a formal hardware/software co-verification approach is presented in Section 5.2.

D. Große and R. Drechsler, *Quality-Driven SystemC Design*,
DOI 10.1007/978-90-481-3631-5_5, © Springer Science+Business Media B.V. 2010

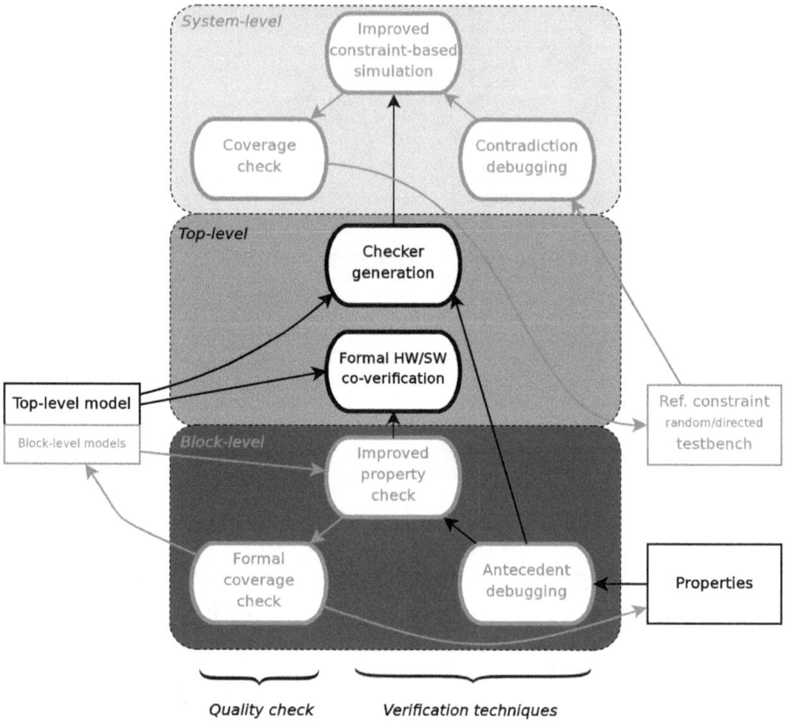

Figure 5.1. Top-level parts of enhanced SystemC design and verification flow

The approach is based on BMC and the formal coverage check to ensure verification quality (see Chapter 4). Besides correctness proofs of the underlying hardware, the hardware/software interface and programs using this interface can be formally verified. The method is shown in Figure 5.1 as the task *formal HW/SW co-verification*. Parts of the results have published in [GKD05a, GKD05b, GKD06].

5.1 Checker Generation

Besides formal verification of the SystemC blocks as described in the previous chapter, their mutual communication has to be checked. This verification task has to be performed at the top level.

There are several approaches for top-level verification which are based on *assertions* (for an overview see e.g. [FKL03]). The key idea is to describe expected or unexpected behavior directly in the device under verification. These conditions are checked dynamically during simulation.

To validate properties of SystemC transaction level models a language extension of *System Verilog Assertions* has been proposed in [EEH+06]. The respective implementation using proxies has been described in [EES+07]

and does not focus on temporal properties. An approach to check temporal assertions for SystemC has been presented in [RHKR01]. There, the specified properties are translated to a special kind of finite state machines (AR-automata). These automata are then checked during the simulation run by algorithms, which have been integrated into the SystemC simulation kernel. In contrast, in [Dre03] a method has been proposed to synthesize properties for circuits into hardware checkers. Properties which have been specified for (formal) verification are directly mapped onto a very regular hardware layout.

Following the latter idea in this section a method is presented which allows checking of temporal properties for circuits and systems described in SystemC not only during simulation. A property is translated into a synthesizable SystemC checker and embedded into the SystemC description. This enables the evaluation of the properties during the simulation and after fabrication of the system. Furthermore, the properties are specified in the standardized *Property Specification Language* (PSL). Of course, with this approach a property is not formally proven and only parts of the functionality are covered. But the proposed method is applicable to large circuits and systems and supports the checking of properties in form of an on-line test. This on-line test is applicable, even if formal approaches failed due to limited resources.

The remaining part of this section is structured as follows: Section 5.1.1 describes the concepts and the translation of a PSL property into a checker circuit. Then, Section 5.1.2 introduces the transformation into a SystemC checker. Experimental results demonstrating the advantages of the approach are given in Section 5.1.3.

5.1.1 Generation of a Checker from a Property

At first the basic idea of the translation of a property into a checker is illustrated by the following example.

EXAMPLE 5.1 *Consider the PSL property shown in Figure 5.2. For the property* `test` *it has to be checked that whenever signal x is* 1, *two time frames later y has to be* 2. *This is equivalent to $\neg(x'' = 1) \vee (y = 2)$, if x'' is x delayed by two clock cycles. If the equation evaluates to false the property is violated. Obviously the translation of the property can be expressed in SystemC*

```
property test = always(
    x == 1
  ) -> (
    next[2](y == 2)
  );
```

Figure 5.2. Example PSL property

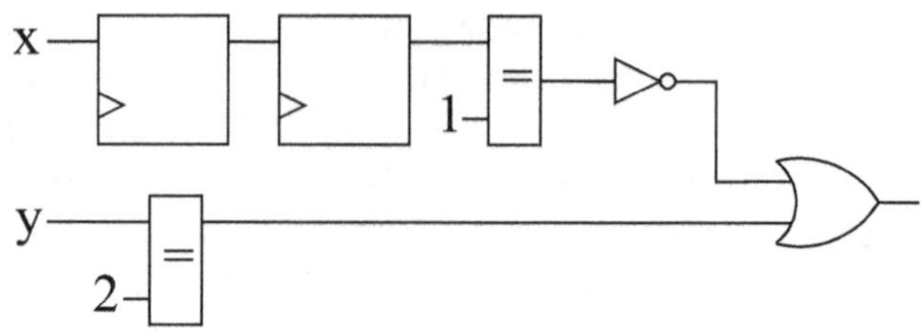

Figure 5.3. Shift register and logic for property `test`

based on shift registers and additional logic representing the relations of the specified signals. The basic idea of a possible hardware realization is shown in Figure 5.3. If the output of the OR gate is 0 *the property fails.*

In general the translation of a property works as follows: Let p be a property which consists of the antecedent A and the consequent C. We separate the antecedent and the consequent into the respective sub-expressions at the logical AND level, i.e. $A = A_1 \wedge A_2 \wedge \ldots \wedge A_m$ and $C = C_1 \wedge C_2 \wedge \ldots \wedge C_n$. Furthermore, we assume that the temporal operators **next_a**[a .. b] and **next_e**[a .. b] are used only as a prefix of A_i or C_j. Then, the translation algorithm is based on four steps:

1. Determine the maximum time point t_{max} of the property by analyzing the time points defined by sub-expressions A_i and C_j. Thereby, also nested utilization of the **next**[i] operator within the temporal operators **next_a**[a .. b] and **next_e**[a .. b] is taken into account.

2. For each signal used in p generate a shift register of length t_{max}. Then, the values of a signal at time points 0, 1, ..., t_{max} are determined by the outputs of the flip-flops in the corresponding shift register. Time point i can directly be identified with the ith flip-flop, if the flip-flops are enumerated in descending order. This is illustrated in Figure 5.4.

3. Combine the signals of each A_i (and C_j) as stated by the logic operations in their expressions. Thereby the variables of the appropriate time points are used. In case of the temporal operators **next_a**[a .. b] and **next_e**[a .. b] an AND and an OR of the resulting expressions is computed, respectively. The results of this step are the equations $\hat{A}_1, \ldots, \hat{A}_m$ and $\hat{C}_1, \ldots, \hat{C}_n$ corresponding to the antecedent and consequent of p, respectively.

4. The final equation for the property p is $check_p = \neg \bigwedge_{i=1}^{m} \hat{A}_i \vee \bigwedge_{j=1}^{n} \hat{C}_j$.

All described transformations from the property into the resulting equation $check_p$ have to be performed by using SystemC constructs, i.e. the use of different data types and operators has to be incorporated. Finally, the property p

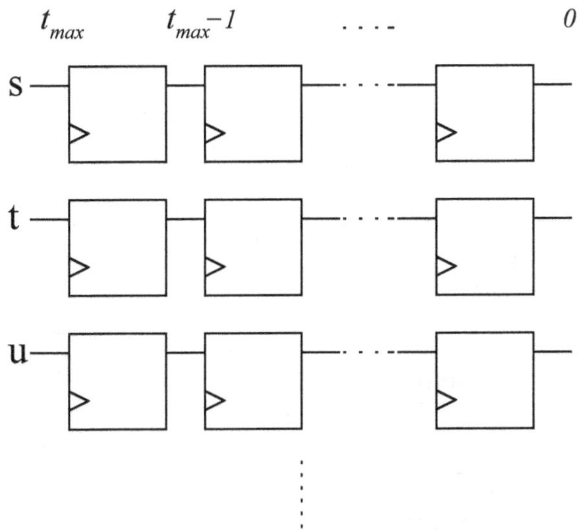

Figure 5.4. Mapping of time points

can be checked by evaluating $check_p$ in each clock cycle during simulation or operation. In the next section the work flow and details about the transformation into SystemC code are given.

5.1.2 Transformation into SystemC Checkers

The work flow of the proposed approach is shown in Figure 5.5. At first the design has to be built and the specification has to be formalized into properties. Then, the properties are translated to checkers and embedded into the design description (shaded area in the figure). If all checkers hold during simulation the next step of the design flow can be entered. Otherwise the design errors have to be fixed or the properties have to be adjusted.

A property is assumed to use only port variables and signals of a fixed SystemC module or from its sub-modules. During the translation for the variables of the properties shift registers have to be created as has been described in the previous section (step 2). For this purpose a generic register as shown in Figure 5.6 has been modeled. The register delays an arbitrary data type for one clock cycle. The data type is specified as a template parameter (see line 1). If such a templated register is not directly supported by the synthesis tool, it is possible to replace every templated register with a version where the concrete input and output types are explicitly specified. The generic register can be used as shown in the example in Figure 5.7. There a register with an `sc_int<8>` input and output is declared and instantiated.

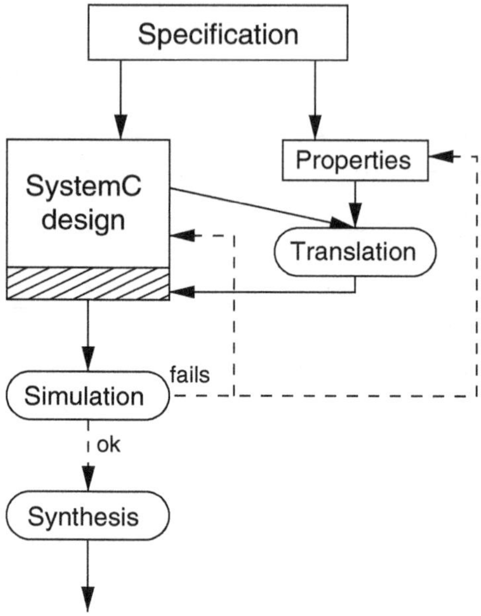

Figure 5.5. Work flow for checker generation

```
template<class T >
class regT : public sc_module {
public:
    sc_in_clk clock;
    sc_in<T > in;
    sc_out<T > out;

    SC_CTOR(regT) {
        SC_METHOD( doit );
        sensitive_pos << clock;
    }

    void doit() {
        out = in.read();
    }
};
```

Figure 5.6. Generic register

During the generation of the shift registers of length t_{max} for a variable, t_{max} generic registers have to be declared and instantiated. This is done in the constructor of the considered module. The necessary `sc_signals` (output variables of the registers) for the different time points are declared as member

```
regT<sc_int<8> > *r = new regT<sc_int<8> >("reg");
r->clock(clock);
r->in(a);
r->out(a_d);
```

Figure 5.7. Usage of generic register

```
SC_MODULE(module) {
public:
    // ports
    sc_in_clk clock;
    ...

    // sc_signals for different time points
    sc_signal<T> x_d1, x_d2;

    SC_CTOR(module) {
        // shift register
        regT<T > *rx_d1 = new regT<T >("reg_rx_d1");
        rx_d1->clock(clock);
        rx_d1->in(x);
        rx_d1->out(x_d1);
        regT<T > *rx_d2 = new regT<T >("reg_rx_d2");
        rx_d2->clock(clock);
        rx_d2->in(x_d1);
        rx_d2->out(x_d2);
        ...
    }
};
```

Figure 5.8. Insertion of a shift register for property `test`

variables of the considered module. Their names are produced by adding the number of delays to the variable name. The absolute time points cannot be used, because if a variable is employed in at least two properties the delay of the same time points may differ.

EXAMPLE 5.2 *Consider again Example 5.1. Let the data type of* `x` *be* `T`. *Let the property* `test` *be written for the SystemC module* `module`. *As has been explained above,* `x` *has to be delayed two times. Then, the resulting shift register is inserted into the module as shown in Figure 5.8.*

As can be seen in Figure 5.8 the data type of a variable used in a property has to be known for declaration of the `sc_signals` and shift registers. Thus, with a simple parser the considered SystemC module is scanned for the data types of the property variables.

```
// property: test
bool check_test = !(x_d2.read() == 1) || (y.read() == 2);
if (check_test == false) {
      cout << "@" << sc_simulation_time();
      cout << ":␣PROPERTY␣test␣FAILS!" << endl;
}
```

Figure 5.9. Checker for property test

The resulting code to check a property (equivalent to equation $check_p$) is embedded into an SC_METHOD process of the module, which is sensitive to the module clock, i.e. the process is triggered every clock cycle. In the final step of generating SystemC code for the translated property the following is taken into account:

- The shift register for each variable used in a property is shared between different checkers.
- In case of an array access it has to be distinguished between an access to an array of ports and an access to a port which contains an array type. An array of ports is mapped onto different variables each representing a different index of the array. Furthermore, the access operator [] has to be replaced accordingly.
- The resulting checker formula is assigned to a Boolean variable check_<property name>. If this variable becomes false during simulation the property is violated and a message is produced using cout. For the synthesis part an output port for the considered module has to be generated, which assumes zero if the property fails.

EXAMPLE 5.3 *In Figure 5.9 the translated equation* check_test *for the property* test *is shown. If the property fails, this is prompted directly to the verification engineer.*

Optimizations

All shift registers for different properties of one concrete module which are driven by the same clock, can be integrated into one clocked process. Then, in the constructor SC_CTOR of the module instead of the shift registers only one clocked process has to be declared. In this process the according output variables are written, e.g. in case of the property test the process statements are:

```
x_d1.write(x); x_d2.write(x_d1);
```

So the number of SC_METHODs is reduced and the simulation speed increases (see also Section 5.1.3).

As has been explained in the previous section if the checkers are synthesized one-to-one for each property an output port is generated, which takes the value zero if the property fails. This leads to a trade-off between good diagnosis and number of output pins. Diagnosis is easy if each property directly corresponds to an output pin, while many output pins require more chip area.

5.1.3 Experimental Results

The technique described above has been experimentally studied. For this task a bus architecture has been modeled. In Figure 5.10 a block diagram of the bus architecture is shown.

The bus is described as a SystemC module, and masters and slaves can connect to the bus. The bus is divided into a data part, an address part and a flag part. These are all `sc_inout`-ports and they are of type `sc_uint` with a scalable size to allow for variable data width, number of slaves, and number of masters. The address is used by the masters to address a slave. The flags `send_flag` and `recv_flag` are set during a bus transaction (see below). Furthermore the bus contains a scalable arbiter. Thus the bus also has a request input and an acknowledge output for each master. The arbiter consists of n cells (one for each master) and combines priority arbitration with a round robin scheme. This guarantees that every master will finally get access to the bus.

Summarized, the features of the bus are

- Only masters can write to the bus and each master has a unique id.
- A slave has a unique address. This address is given at instantiation of the slave.

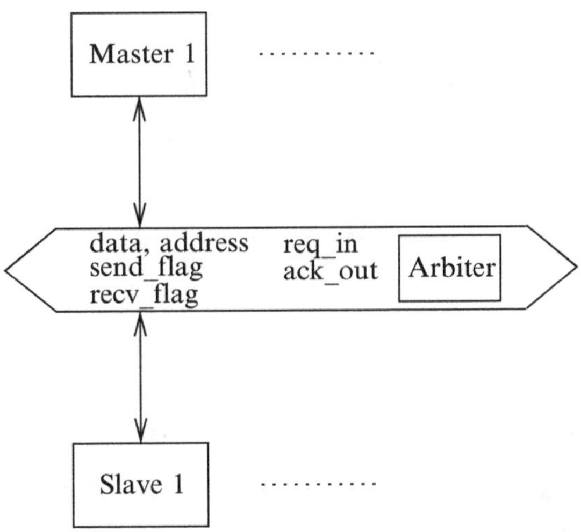

Figure 5.10. Bus architecture

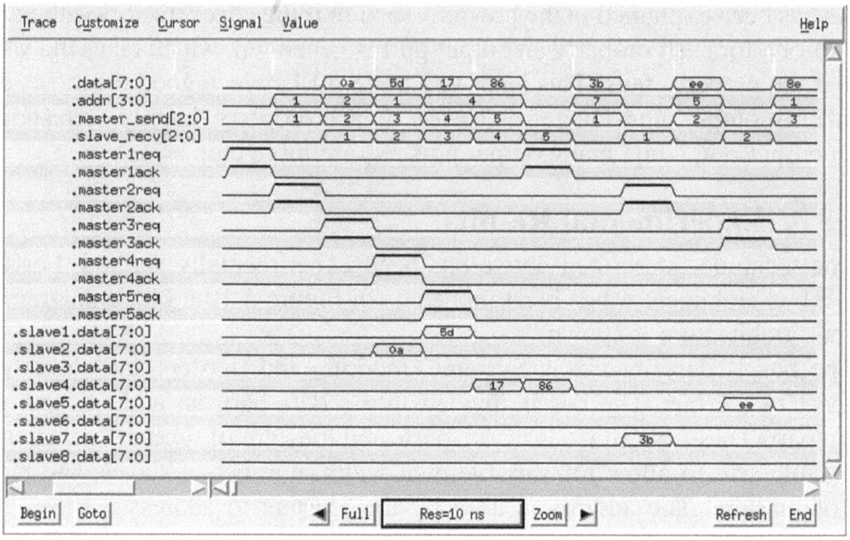

Figure 5.11. Simulation trace of a bus example

- A bus transaction works as follows:

  1. A master requests the bus via its request output. If access is granted see step 2, otherwise the master waits for an acknowledgment.
  2. The master writes the target address and the data to the bus. Furthermore, the master writes its id to the `send_flag`. Then, the master waits for an acknowledge that the slave has received the data via the `recv_flag` (id of the master at the `recv_flag`).
  3. If a slave detects its address on the bus, the slave reads the data and writes the id from the `send_flag` to the `recv_flag` of the bus.
  4. If the master detects its id on the bus, the data transmission was successful.

A waveform example of a bus with five masters and eight slaves is shown in Figure 5.11.

Checkers

In the following an informal description of the properties is given, that have been embedded as checkers into the bus module:

1. Two output signals of the arbiter can never become 1 at the same time (*mutual exclusion*).

2. The acknowledge is only set if there has been a request (*conservativeness*).

3. Each request is confirmed by an acknowledge within $2 \cdot n$ time frames (*liveness*).

4. If the bus has been granted for a master, the master writes its id to the `send_flag` in the next cycle (*master id*).

5. If a slave has been addressed, the slave writes the master id (available at the `send_flag`) to the `recv_flag` (*acknowledge master*).

Simulation Results

All experiments have been carried out on an Intel Pentium IV 3 GHz with 1 GB of main memory running Linux. Checkers have been generated for all described properties (see previous section). In the following for each property the simulation performance in case of no checkers, the simple approach, and the optimized approach are compared. For this task the bus model has been simulated for 100,000 clock cycles for a different number of masters. Note that the number of masters connected to the bus is equal to the number of arbiter cells. For the checkers described above we obtained the following results:

1. In Figure 5.12 the performance comparison for the checker *mutual exclusion* is shown. As can be seen the simulation time for the simple and the optimized approach increases with the number of masters. Both approaches behave similar since the property interval of the mutual exclusion property considers only one time point, so no registers have to be created. For this reason no optimization is possible. The total run-time overhead of up

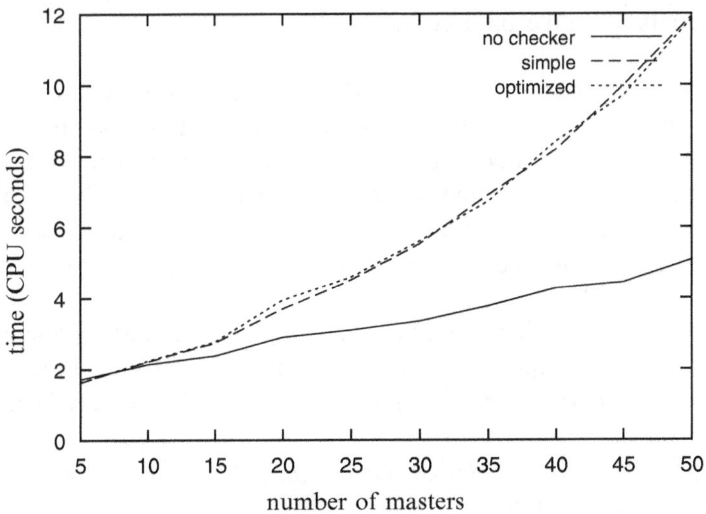

Figure 5.12. Comparison of simulation performance for checker *mutual exclusion*

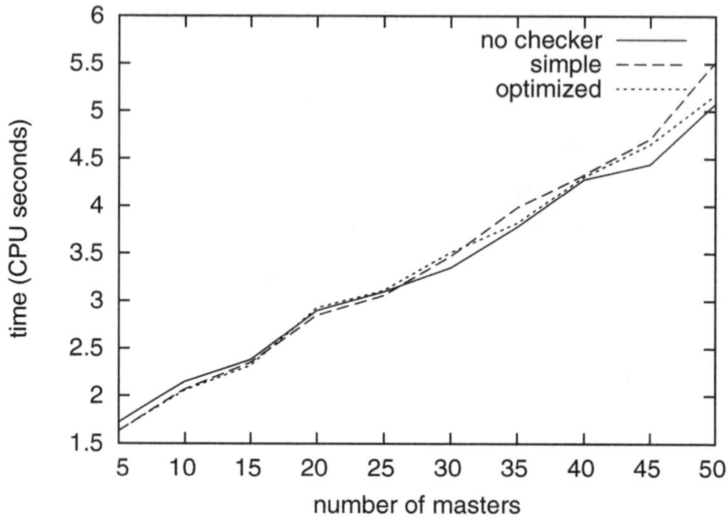

Figure 5.13. Comparison of simulation performance for checker *conservativeness*

to the factor of 2.5 in Figure 5.12 is due to the quadratic nature of the mutual exclusion property, i.e. each pair of distinct acknowledge signals is accessed in the property.

2. The simulation performance with and without the checkers for the *conservativeness* properties is nearly identical (see Figure 5.13). This is an expected behavior, because each conservativeness property only argues over two signals of each arbiter cell.

3. In Figure 5.14 the results for the *liveness* checkers are depicted. The runtime overhead compared to pure simulation is due to the significantly increasing size of the time interval of the liveness properties, i.e. the size is $2 \cdot n$, where n is the number of masters. As can be seen the optimized approach leads to better results than the simple approach because the number of `SC_METHOD`s has been reduced effectively by optimization.

4. The results for the checkers of *master id* show that there is a small benefit of the optimized approach over the simple approach (see Figure 5.15). In total these properties can be checked during simulation very fast.

5. As expected the *acknowledge master* property leads to the same performance as pure simulation, because this property only relates two time points, i.e. the time interval of the property is $[0, 1]$. Figure 5.16 shows the diagram.

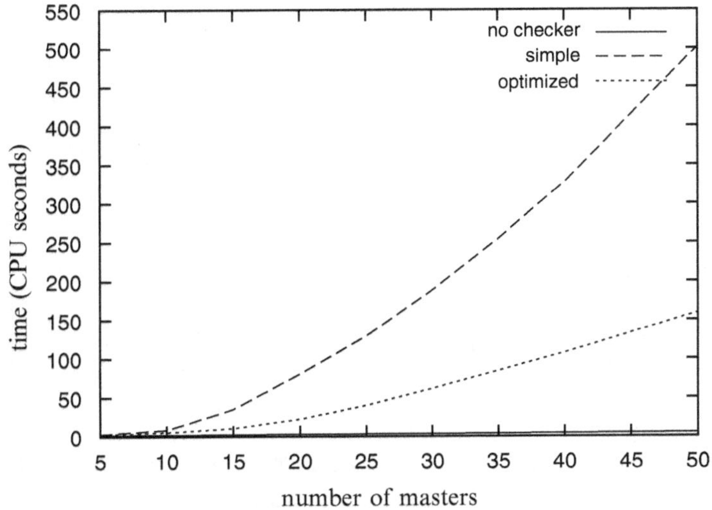

Figure 5.14. Comparison of simulation performance for checker *liveness*

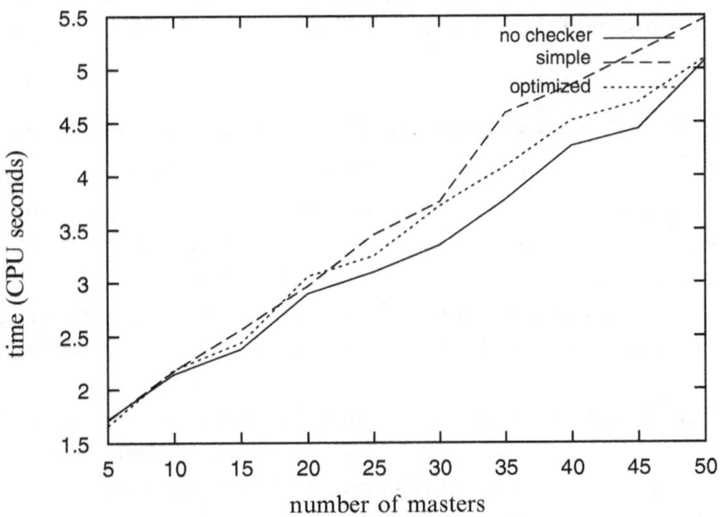

Figure 5.15. Comparison of simulation performance for checker *master id*

The experiments demonstrate that the overhead during simulation for properties with large time intervals is acceptable, and negligible for properties with smaller time intervals.

Overall, the presented approach enables the verification of complex communication of blocks at the top level by embedding checkers directly into

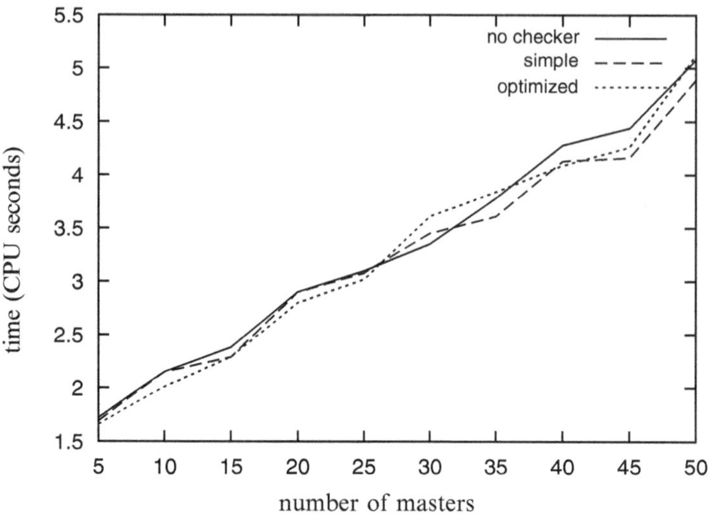

Figure 5.16. Comparison of simulation performance for checker *acknowledge master*

the SystemC description. In the next part of this chapter a formal verification approach is presented that takes the characteristics of embedded systems into account.

5.2 HW/SW Co-Verification for Embedded Systems

In the last few years embedded system design has become a very important research area and the application domains range from telecommunication devices to automotive units. These systems not only consist of hardware components, i.e. a large portion is realized by firmware and programs. Since these systems are used more and more in safety-critical applications, the aspect of verification is very important to ensure the correct functional behavior of the system.

In the meantime hardware verification has been intensively studied and successfully used, even though the tools sometimes suffer from limit of resources. But assertion-based verification and formal approaches have ensured high quality also for large hardware systems. This standard so far is not achieved, if software components[1] are included. For example, a study by Collett International Research Inc. has shown that errors caused by firmware and hardware/software interfaces account for up to 13of failures with an increasing trend. To reduce this type of errors in embedded systems integrated hardware/software verification is needed.

[1]In this section we consider hardware dependent software, i.e. there is a direct software to hardware correspondence.

A successful technique for verification of hardware is *Bounded Model Checking* (BMC) [BCCZ99] (see also Section 2.3.2). BMC checks whether a circuit satisfies a temporal property or not. Therefore, BMC reduces the verification problem to a *Boolean Satisfiability* (SAT) problem and searches for counter-examples in executions whose length is bounded by k time steps.

In this section we show that the concepts of BMC[2] can also be applied in the context of hardware/software integration. For an embedded system, that in our context consists of digital components without analog units, complete verification can be performed. This includes the underlying hardware, interface instructions and programs based on sequences of instructions. Arguing over the behavior of a program becomes possible by specifying the corresponding sequences of instructions as assumptions in the antecedent of the BMC property and formulating the intended behavior as goal in the consequent. In total, the integrated hardware/software verification approach allows the complete formal verification of an embedded system.

The remaining part of this section is structured as follows. We introduce the notion of a *Timed Embedded System* (TES) that defines the type of embedded system that can be verified by our approach in Section 5.2.1. Then, the integrated verification approach is presented in Section 5.2.2. The verification of a TES is divided into hardware, interface and program verification. These steps are discussed and in each phase the application of BMC is explained. A case study demonstrates the verification of a RISC CPU in Section 5.2.3. Following the approach, first, the correctness of the underlying hardware is shown. In the second step the hardware/software interface of the RISC CPU is formally verified. Based on this result, assembler programs for the RISC CPU are considered and successfully verified.

5.2.1 Co-Verification Model

Before the details on the integrated verification approach for hardware and software are given we specify the type of embedded systems that is considered. Then, the verification steps that allow the formal co-verification of an embedded system are explained.

Timed Embedded System

In the following we restrict ourselves to embedded systems that guarantee a response in a fixed number of cycles. We call such a system a *Timed Embedded System* (TES). These systems include all kinds of digital microprocessors, e.g. specialized DSPs or RISC CPUs.

[2]See Section 4.1 for BMC variant used here.

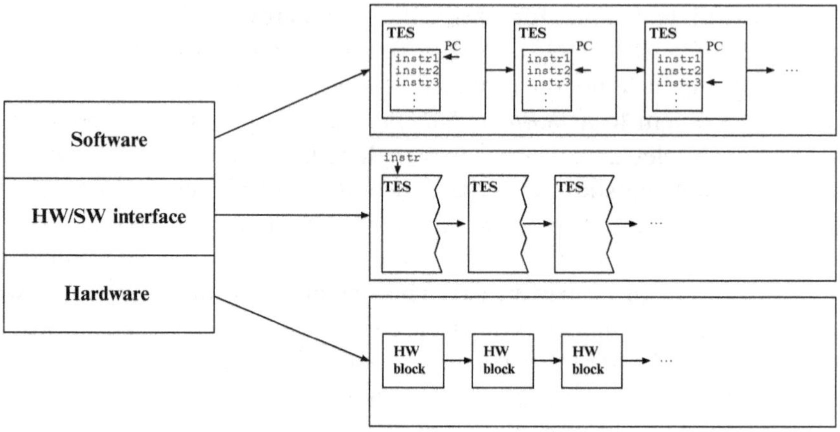

Figure 5.17. TES architecture and models for verification

A simplified architecture of a TES is shown in the left part of Figure 5.17. The hardware layer includes hardware blocks like memories, ALUs, etc. The hardware/software interface layer defines ports and instructions for the communication between software and the underlying hardware. On top, the software layer contains the programs.

Often in the area of test or hardware verification as underlying model the unrolled circuit is used. We apply this concept for the verification of a TES. But here not only hardware is verified. In fact, the correctness of software can be shown as well. Software of a TES consists of instructions that access the underlying hardware via an interface. This allows to consider hardware, interface and software in one integrated system view model.

In the following section these observations are explained in more detail.

5.2.2 Co-Verification Steps

Hardware

To verify the underlying hardware we directly apply BMC for all hardware units. For each block several temporal properties are proven. The model for formal verification of a hardware block is shown in the lower right part of Figure 5.17. As illustrated a block is unrolled up to the size of the time interval which depends on the timing constructs used in a property. In total the functional correctness of each hardware block is verified. The completeness of the formal verification of each hardware block is shown with the approach presented in Section 4.4. Thus, the properties form an unambiguous specification of each hardware block. The results of this verification step are the basis for the interface verification.

Interface

The interface is viewed as a specification that exists between hardware and software. By calling instructions of an interface, programs can communicate with the underlying hardware. At the interface the functionality of the hardware is available but the concrete hardware realization is hidden. In contrast to hardware verification, the interface verification with BMC formulates for each interface instruction the exact response of all hardware blocks involved. Besides this, it is also assured that no side effects occur. The unrolled model only consists of parts of the TES. In particular, in each property the considered interface instruction is assumed to be executed in the first cycle (see right hand side of Figure 5.17, middle). The objective of interface verification is to guarantee the correctness of all interface instructions. Therefore, again the formal coverage check as introduced in the previous chapter is used. Hence, all interface instructions have to be proven since otherwise full coverage cannot be achieved for the state holding elements at the top level which are updated by the interface instructions. Overall, this step forms the basis for program verification.

Program

Based on instructions available at an interface a program is a structural sequence of instructions. By a combination of BMC and inductive proofs [BC00, SSS00] a concrete program can be formally verified. Arguing over the behavior of a program is possible by constraining the considered sequence of instructions as assumptions in the antecedent of a BMC property. Thus, the property checker "executes" the program and can check the intended behavior in the consequent of the property. The intended behavior specified in the consequent of the property, requires a careful consideration of the effects of the used instructions in a program. For different example programs this is detailed later. As a model for program verification, the TES is unrolled and the instructions of the program under verification are assumed to be executed next. Therefore, in the antecedent of the property it is constrained that the instructions reside in the memory and the program counter points to the first instruction. The upper right part of Figure 5.17 illustrates this procedure. Inductive reasoning is used to verify properties which describe functionality where the upper time bound varies, e.g. this can be the case if loops are used.

In the following section the integrated verification approach is applied to a RISC CPU.

5.2.3 Experimental Results

This section provides the basics of the RISC CPU and the SystemC model of this CPU. Afterwards, the proposed integrated verification approach is applied to the RISC CPU and some example programs. All experiments have been carried out on an AMD Athlon XP 2800+ with 1 GB of main memory.

Specification and SystemC Model

In Figure 5.18 the main components of the RISC CPU are shown. We use the same RISC CPU as in Section 4.4.3. The basic data of the RISC CPU is briefly reviewed. In addition, more details with respect to the instructions are provided.

The CPU has been designed as a Harvard architecture. The data width of the program memory and the data memory is 16 bit. The size of the program memory is 4 kByte and the size of the data memory is 128 kByte. The length of an instruction is 16 bit. We only briefly describe the five different classes of instructions in the following:

- 6 load/store instructions (movement of data between register bank and data memory or I/O device, loading of a constant into high- or low-byte of register)
- 8 arithmetic instructions (addition/subtraction with and without carry, left/ right rotation and shift)
- 8 logic instructions (bit by bit negation, bit by bit exor, conjunction/ disjunction of two operands, masking, inverting, clearing and setting of single bits of an operand)
- 5 jump instructions (unconditional jump, conditional jump, jump on set/ cleared carry or zero flag)
- 5 other instructions (stack instructions push and pop, program halt, subroutine call, return from subroutine)

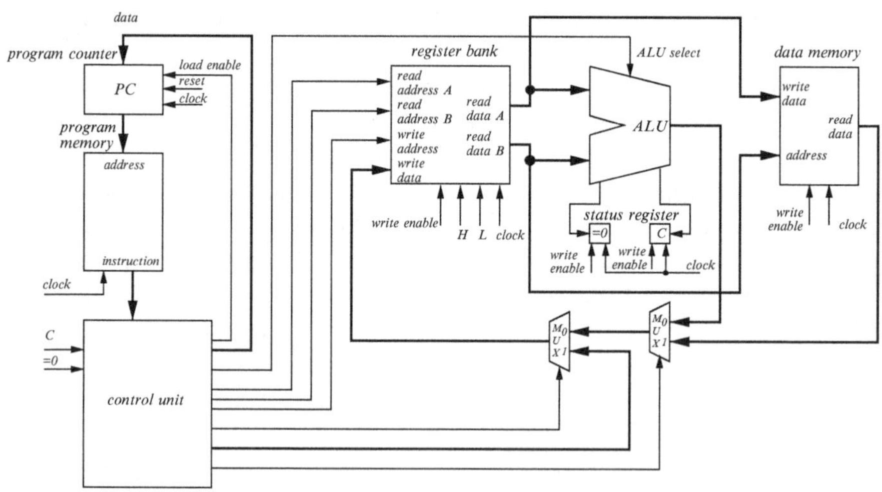

Figure 5.18. Structure of the RISC CPU including data and instruction memory

Since we later consider different assembler programs a short overview including the assembler notation for each instruction is provided in Table 5.1. For more details on the CPU specification we refer the reader to [BDM05]. The RISC CPU has been modeled in SystemC. Details on the SystemC model can be found in [GKG$^+$05, Kue06]. For the RISC CPU a compiler has been implemented which generates object code from an assembler program. This object code runs on the SystemC model, i.e. the model of the CPU executes an assembler program.

Table 5.1: Instructions of RISC CPU

Assembler code	**Description**
Load/store instructions	
LDD $R[i]$, $R[j]$	Load memory content from address $R[j]$ into $R[i]$
STO $R[j]$, $R[k]$	Store $R[k]$ into memory at address $R[j]$
LDL $R[i]$, d	Load constant d into low part of $R[i]$
LDH $R[i]$, d	Load constant d into high part of $R[i]$
IN $R[i]$, $R[j]$	Load data from I/O device address $R[j]$ into $R[i]$
OUT $R[j]$, $R[k]$	Save data of $R[k]$ to I/O device address $R[j]$
Arithmetic instructions	
ADC $R[i]$, $R[j]$, $R[k]$	Addition with carry into $R[i]$
SBC $R[i]$, $R[j]$, $R[k]$	Substraction with carry into $R[i]$
ADD $R[i]$, $R[j]$, $R[k]$	Addition w/o carry into $R[i]$
SUB $R[i]$, $R[j]$, $R[k]$	Substraction w/o carry into $R[i]$
ROR $R[i]$, $R[j]$	Bitrotation right of $R[j]$
ROL $R[i]$, $R[j]$	Bitrotation left of $R[j]$
SHR $R[i]$, $R[j]$	Bitshift right of $R[j]$
SHL $R[i]$, $R[j]$	Bitshift left of $R[j]$
Logic instructions	
NOT $R[i]$, $R[j]$	Bit by bit negation
XOR $R[i]$, $R[j]$, $R[k]$	Bit by bit exor
OR $R[i]$, $R[j]$, $R[k]$	Bit by bit or
AND $R[i]$, $R[j]$, $R[k]$	Bit by bit and
MKB $R[i]$, $R[j]$, b	Masking of bit b
INB $R[i]$, $R[j]$, b	Inverting of bit b
SEB $R[i]$, $R[j]$, b	Set bit b
CLB $R[i]$, $R[j]$, b	Clear bit b
Jump instructions	
JMP d	Jump to address d
JC d	Jump to address d, if carry is set

`JZ` d	Jump to address d, if zero-flag is set
`JNC` d	Jump to address d, if carry is not set
`JNZ` d	Jump to address d, if zero-flag is not set
Other instructions	
`PSH` $R[k]$	Push $R[k]$ to stack
`POP` $R[i]$	Pop from stack into $R[i]$
`JS` d	Jump to subroutine at address d
`RTS`	Return from subroutine
`HLT`	Program halt

Table 5.2. Results for hardware verification

Block	#p	CPU time (s)	Max. mem (MB)
ALU	18	4.30	15
Program memory	2	1.21	21
Data memory	2	4.52	41
Register bank	5	1.22	15
Program counter	4	0.10	8
Stack pointer	6	0.09	8
Control unit	19	0.23	8

Formal Co-Verification

For BMC of the RISC CPU the approach presented in Section 4.1 is used. In the following the complete verification of the hardware, interface and programs for the RISC CPU is discussed.

Hardware. Properties for each block of the RISC CPU have been formulated. For example, for the control unit it has been verified which control lines are set according to the opcode of the instruction input. Overall the correctness of each block has been verified. Note that based on the functional coverage approach for BMC presented in the previous chapter 100% coverage of each hardware block is guaranteed (see Section 4.4.3). Table 5.2 summarizes the results. The first column gives the name of the considered block. Next, the number of properties specified for a block are denoted. The last columns provide the overall run-time needed to prove all properties of a block and the maximal used memory during verification, respectively. As can be seen the functional correctness of the hardware blocks has been formally verified very fast with 50 properties. The required effort to ensure full coverage for each hardware block is in the same order of magnitude (see Section 4.4.3 for the details).

Interface. Based on the complete formal hardware verification of the RISC CPU, in the next step the interface is verified. Thus, for each instruction of the RISC CPU a property has been specified which expresses the effects on all hardware blocks involved. Also for the verification of each instruction at the top level of the RISC CPU the formal coverage approach from Chapter 4 has been applied (see Section 4.4.3) and hence full functional coverage was achieved here. The fundamental verification task at the interface is illustrated by the following example.

EXAMPLE 5.4 *Figure 5.19 gives details on the ADD instruction. Besides the assembler notation also the instruction format of the ADD instruction is depicted. The symbol* $bin(i)$ *denotes the binary encoding of the natural number* i. *The specified property for the ADD instruction is shown in Figure 5.20. First of all the opcode and the three addresses of the registers are assigned to variables (lines 1–6) which make the following property expressions easier to read. The antecedent of the ADD property is defined from line 11 to 12 and states that there is no reset (line 11), the current instruction is addition (line 12) and the registers* R[0] *and* R[1] *are not addressed (since these registers are special purpose registers that contain the constants zero and one, respectively). Under these assumptions we prove that in the next cycle the register* R[i] *(=reg . reg[**prev**(Ri_A)]) contains the sum of register* R[j] *and register* R[k] *(line 17), the carry (stat . C) in the status register is updated properly (line 16) and the zero bit (stat . Z) is set if the result of the sum is zero (line 18). Furthermore, we prove that the ADD instruction has no side effects, i.e. the contents of all registers which are different from* R[i] *remain unchanged.*

Analogously to the ADD instruction the complete instruction set of the RISC CPU is verified. Table 5.3 summarizes the results.

The first column gives the category of the instruction. In the second column the number of properties for each category is provided. The last two columns show the total run-time needed to prove all properties of a category and the maximum memory needed during verification, respectively. As can be seen

Assembler notation: ADD $R[i]$,$R[j]$,$R[k]$

Task: addition of $R[j]$ and $R[k]$, the result is stored in $R[i]$

Instruction format:

15	…			11	10 9 8	7 6	5 4 3	2 1 0
0	0	1	1	1	$bin(i)$	- -	$bin(j)$	$bin(k)$

Figure 5.19. ADD instruction

```
assign OPCODE = instr.range(15,11);
assign Ri_A = instr.range(10,8);
assign Rj_A = instr.range(5,3);
assign Rk_A = instr.range(2,0);
assign Rj = reg.reg[Rj_A];
assign Rk = reg.reg[Rk_A];

property ADD =
always(
 // antecedent
  reset == 0 && OPCODE == "00111" &&
  Ri_A > 1 && Rj_A > 1 && Rk_A > 1
 ) -> (
 // consequent
  next(
   (reg.reg[prev(Ri_A)] + (65536 * stat.C) == prev(Rj) + prev(Rk))
    && ((reg.reg[prev(Ri_A)] == 0) ? (stat.Z == 1) : (stat.Z == 0))

   // no side effects
   && ( (prev(Ri_A) != 2) ? (reg.reg[2] == prev(reg.reg[2])) : 1 )
   && ( (prev(Ri_A) != 3) ? (reg.reg[3] == prev(reg.reg[3])) : 1 )
   && ( (prev(Ri_A) != 4) ? (reg.reg[4] == prev(reg.reg[4])) : 1 )
   ...
  )
);
```

Figure 5.20. Specified property for the ADD instruction of the RISC CPU

Table 5.3. Run-time of interface verification

Category	#p	CPU time (s)	Max. mem (MB)
Load/store	6	79.70	93
Arithmetic	8	970.94	216
Logical	8	47.76	79
Jump	5	13.83	69
Other, reset	6	26.02	76

the complete instruction set of the RISC CPU can be verified in less than 20 CPU minutes. For details on the coverage analysis see Section 4.4.3.

Program. Finally, we describe the approach to verify assembler programs for the RISC CPU. As explained, the considered programs of the RISC CPU can be verified by constraining the instructions of the program as assumptions in the proof. These assumptions are automatically generated by the compiler of the RISC CPU. The verification of programs is illustrated by three case studies.

Loop Unrolling. Consider the assembler program shown in Figure 5.21. The program loads the integer 10 into register $R[7]$ and decrements register $R[7]$ in a loop until it contains value 0. For this program the property `count` has been formulated (see Figure 5.22). At first it is assumed that the CPU memory contains the instructions of the given example (lines 4–7).[3] Furthermore, the program counter points to the corresponding memory position (line 8), no memory write operation is allowed (line 9) and there is no reset for the considered 22 cycles (line 10). Then, we prove that register $R[7]$ is zero after 21 cycles (line 13). The time point 21 results from the fact that the first two cycles (zero and one) are used by the load instructions and the following 20 cycles are required to loop 10 times. The complete proof has been carried out in less than 25 CPU seconds.

Fibonacci Numbers. An assembler program has been written that computes the Fibonacci numbers (defined as $f(n) = f(n-1) + f(n-2)$ with

```
/* counts from 10 down to 0  */
          LDL R[7], 10
          LDH R[7], 0
loop:
          SUB R[7], R[7], R[1]
          JNZ loop
```

Figure 5.21. Example assembler program

```
1  property count =
2  always(
3    // antecedent
4    rom.mem[0] == 18186 &&   // LDL R[7], 10
5    rom.mem[1] == 20224 &&   // LDH R[7], 0
6    rom.mem[2] == 14137 &&   // SUB R[7], R[7], R[1]
7    rom.mem[3] == 24578 &&   // JNZ 2
8    pc.pc == 0 &&
9    next_a[0..21](prog_mem_we == 0) &&
10   next_a[0..21](reset == 0)
11 ) -> (
12   // consequent
13   next[21] (reg.reg[7] == 0 )
14 );
```

Figure 5.22. Property `count`

[3]This part of the assumptions has been generated automatically by the compiler.

$f(0) = 1$ and $f(1) = 1$). We only give the results. The correctness of the program has been verified by induction. In the property for the base case the result for $f(0)$ and $f(1)$ has been proven. The induction step formulates that in each loop the next Fibonacci number is computed by adding the two previous Fibonacci numbers. In total the correctness of the Fibonacci program has been proven in less than 20 CPU seconds.

Multiplication. Since the ALU of the RISC CPU has no multiply operation this functionality has to be implemented as a program. Figure 5.23 shows a program that performs an 8-bit multiplication. The program is based on shift instructions and addition instructions. In the beginning the two multiplication factors are in registers $R[2]$ and $R[3]$. The partial product is kept in $R[5]$ during multiplication. The result is stored in $R[6]$. Register $R[7]$ is used as a counter and is initialized to 8 in lines 6 and 7. In the loop, the instruction in line 9 tests the next bit in the first factor. If the bit is set, the current partial product is added to the result (line 11). The shift instruction in line 13 computes the next partial product. Then the counter is decremented (line 14) and the loop continues until $R[7]$ reaches the value 0. Note that the number of cycles needed to complete the program depends on the number of bits set to 1 in the first factor because line 11 may not be executed in every loop.

Figure 5.24 shows a PSL property for the described assembler program. At first the multiplication factors in registers $R[2]$ and $R[3]$ are assigned to the variables `FAC1` and `FAC2`. As assumption it is required that the multiplication program is located in the memory (lines 6–10). Again, this part has been generated automatically. Line 11 states that the program counter points to the first

```
1  mult:
2          OR  R[4], R[2], R[0]
3          OR  R[5], R[3], R[0]
4          LDL R[6], 0
5          LDH R[6], 0
6          LDL R[7], 8
7          LDH R[7], 0
8  loop:
9          SHR R[4], R[4]
10         JNC 11
11         ADD R[6], R[6], R[5]
12 11:
13         SHL R[5], R[5]
14         SUB R[7], R[7], R[1]
15         JNZ loop
```

Figure 5.23. Assembler program for 8-bit multiplication

```
1  assign FAC1 = reg.reg[2];
2  assign FAC2 = reg.reg[3];
3  property mul =
4  always(
5    // antecedent
6    rom.mem[0] == 5136   &&  // OR R[4], R[2], R[0]
7    rom.mem[1] == 5400   &&  // OR R[5], R[3], R[0]
8    rom.mem[2] == 17920  &&  // LDL R[6], 0
9    ...
10   rom.mem[11] == 24582 &&  // JNZ 6
11   pc.pc == 0 &&
12   next_a[0..54](prog_mem_we == 0) &&
13   next_a[0..54](reset == 0) &&
14   FAC1 < 256 && FAC2 < 256
15 ) -> (
16   // consequent
17   next_e[46..54] (pc.pc == 12) &&
18        (reg.reg[6] == FAC1 * FAC2) );
```

Figure 5.24. Property `mul`

instruction of the multiplication program in the beginning. Lines 12 and 13 assure that no reset and no write access to the memory take place during program execution. Finally, line 14 requires that both multiplication factors are within an 8-bit range. Under these assumptions we prove that between cycles 46 and 54 after starting the algorithm[4] the program counter points to the next instruction[5] (line 17) and register $R[6]$ contains the product at the same time (line 18). In other words, we prove that the assembler program does in fact perform a multiplication.

The SAT instance generated for this property consisted of $2,894,173$ clauses and $6,735,707$ literals. The correctness of the multiplication program has been verified in less than $4,000$ CPU seconds.

To summarize, the proposed approach allows the hardware/software co-verification of timed embedded systems. As has been shown the correctness of simple and complex programs has been proven.

[4]As mentioned above, the number of cycles depends on the input. Six cycles are needed for the first six instructions plus eight times the loop of five or six instructions.
[5]This instruction is the first after the instruction JNZ loop.

5.3 Summary and Future Work

We presented two techniques for verifying the functionality at the top level. Since the entire top-level model cannot be handled completely by formal methods an approach to generate checkers from temporal properties has been proposed. The resulting SystemC checkers are validated during simulation and can be also synthesized for on-line tests after fabrication. The simulation is performed using the improved system-level techniques and hence the system-level quality check is also available after refining the system. For a bus architecture the benefits of the approach have been demonstrated.

Then, a formal hardware/software co-verification approach for embedded systems has been presented. The approach is based on BMC and the formal coverage check as introduced in the previous chapter. On top of hardware verification the correctness of the interface is proven. In the last step software which uses this interface can be formally verified which has been demonstrated for different examples.

Possible directions for future work are the combination of the system-level coverage check with the checker generation approach. The idea is to ensure that the antecedent of each checker has to evaluate to true at least once. For the hardware/software co-verification method we plan to automatically derive time bounds from assembler programs to aid the specification of the consequent of software properties. Also the formal coverage analysis technique proposed in the previous chapter has to be extended to check the coverage of software properties.

In summary, the techniques presented in this chapter allow a continuous design and verification flow by connecting the system level and the block level. Based on this tight integration of the dedicated verification techniques and the complementing coverage checks a high quality system can be build since the verification quality is guaranteed along the complete design flow from abstract descriptions down to synthesizable ones.

Chapter 6

SUMMARY AND CONCLUSIONS

Today, the design and verification of electronic systems is a very challenging task. To cope with the steadily increasing complexity and the pressure of time-to-market the design entry has been lifted to high-level descriptions, i.e. the level of abstraction for designing systems has been raised. In this book a prominent design flow based on the system description language SystemC was considered. However, while the single-language concept of the SystemC design flow allows a continuous modeling from system level down to synthesizable descriptions, only low verification quality is achieved. There are two main reasons: First, the existing verification techniques are decoupled and are often based on simple simulation techniques. Second, the resulting verification quality in terms of the covered functionality is not ensured automatically along the refinement process. Therefore, a *quality-driven design and verification flow* was developed in this book. The "traditional" SystemC design flow is enhanced by

1. dedicated verification techniques which target each level of abstraction and employ formal methods where possible and

2. complementing each verification task by measuring the resulting quality.

In the new flow three levels of abstraction for modeling of digital systems are distinguished: The system-level model that is refined to the synthesizable top-level model which again consists of several block-level models.

For verifying the system-level model "pure" simulation is replaced by improved constraint-based simulation in the new flow. Hence, test scenarios are generated that satisfy given constraints. In particular corner-case scenarios are produced very effectively. The SystemC-based implementation on top of the SCV library is improved such that new operators are available as well

D. Große and R. Drechsler, *Quality-Driven SystemC Design*,
DOI 10.1007/978-90-481-3631-5_6, © Springer Science+Business Media B.V. 2010

as a uniform distribution among all constraint solutions is guaranteed. Since over-constraining occurs while formulating specific constraints to check complex behavior an approach for debugging was proposed. The method finds all minimal reasons of the constraint contradictions and hence the debugging time is reduced significantly. The resulting verification quality is checked by an approach which measures how thorough the system was tested. In a coverage report the results are presented. Thus, an automatic quality check is guaranteed.

In the next step, the block-level verification starts since the correctness of the block-level models can be proven. For this task a SystemC property checker using bounded model checking was developed. The temporal properties are formulated in the standardized property language PSL as implications. The property checker is improved by reusing learned information while weakening the antecedent of a property; this results in faster proofs. In addition, an approach for debugging of a contradictory property was presented which identifies automatically whether a contradiction results solely form the property or in combination with the design. For ensuring high verification quality an approach to analyze the coverage of property checking was developed. By closing all coverage gaps which are automatically identified by the approach, it is guaranteed that the behavior of each block is specified unambiguously by the properties. Hence, here the highest possible quality is achieved.

Based on the block-level verification results the top level is considered. There, an approach to generate checkers from temporal properties was proposed which are validated during simulation. Thus, the verification quality is ensured with the respective system-level method. Furthermore, for embedded systems a complete formal hardware/software co-verification methodology was presented. Besides the hardware and the interface also programs are formally verified.

In several experiments the proposed techniques were evaluated using different benchmarks and have shown to be very effective. At the end of each chapter several directions and ideas for future work were discussed. In summary, the new design and verification flow allows to develop high quality systems. This becomes possible due to specialized verification techniques which heavily use formal methods as well as consequently checking the verification quality in an automatic way. Overall, following the introduced flow helps design teams and verification engineers in delivering correct systems which is important for consumer electronics and life-saving in case of safety-critical devices.

References

[Acc05] *Accellera Property Specification Language Reference Manual, version 1.1.* http://www.pslsugar.org, 2005.

[ADK+05] N. Amla, X. Du, A. Kuehlmann, R. P. Kurshan, and K. L. McMillan. An analysis of SAT-based model checking techniques in an industrial environment. In *Correct Hardware Design and Verification Methods*, pages 254–268, 2005.

[AFF+03] R. Armoni, L. Fix, A. Flaisher, O. Grumberg, N. Piterman, A. Tiemeyer, and M. Y. Vardi. Enhanced vacuity detection in linear temporal logic. In *Computer Aided Verification*, volume 2725 of *LNCS*, pages 368–380. Springer, 2003.

[AKMM03] N. Amla, R. P. Kurshan, K. L. McMillan, and R. Medel. Experimental analysis of different techniques for bounded model checking. In *Tools and Algorithms for the Construction and Analysis of Systems*, pages 34–48, 2003.

[AVS+04] M. F. Ali, A. Veneris, S. Safarpour, R. Drechsler, A. Smith, and M. S. Abadir. Debugging sequential circuits using Boolean satisfiability. In *Int'l Conf. on CAD*, pages 204–209, 2004.

[Bai04] B. Bailey. A new vision of scalable verification. *EE Times*, 2004. http://www.eetimes.com/showArticle.jhtml?articleID=18400907.

[BAMP81] M. Ben-Ari, Z. Manna, and A. Pnueli. The temporal logic of branching time. In *Symposium on Principles of Programming Languages*, pages 164–176, 1981.

[BAS02] A. Biere, C. Artho, and V. Schuppan. Liveness checking as safety checking. In *International Workshop on Formal Methods for Industrial Critical Systems*, 2002.

[BB94] D. L. Beatty and R. E. Bryant. Formally verifying a microprocessor using a simulation methodology. In *Design Automation Conf.*, 1994.

[BBDER97] I. Beer, S. Ben-David, U. Eisner, and Y. Rodeh. Efficient detection of vacuity in ACTL formulas. In *Computer Aided Verification*, volume 1254 of *LNCS*, pages 279–290, 1997.

[BBDER01] I. Beer, S. Ben-David, C. Eisner, and Y. Rodeh. Efficient detection of vacuity in temporal model checking. *Formal Methods in System Design*, 18(2):141–163, 2001.

[BBM+07] J. Bormann, S. Beyer, A. Maggiore, M. Siegel, S. Skalberg, T. Blackmore, and F. Bruno. Complete formal verification of Tricore2 and other processors. In *Design and Verification Conference (DVCon)*, 2007.

[BC00] P. Bjesse and K. Claessen. SAT-based verification without state space traversal. In *Int'l Conf. on Formal Methods in CAD*, volume 1954 of *LNCS*, pages 372–389. Springer, 2000.

[BCC+03] A. Biere, A. Cimatti, E. M. Clarke, O. Strichman, and Y. Zhu. Bounded model checking. *Advances in Computers*, 58:118–149, 2003.

[BCCZ99] A. Biere, A. Cimatti, E. M. Clarke, and Y. Zhu. Symbolic model checking without BDDs. In *Tools and Algorithms for the Construction and Analysis of Systems*, volume 1579 of *LNCS*, pages 193–207. Springer Verlag, 1999.

[BCL+94] J. R. Burch, E. M. Clarke, D. E. Long, K. L. McMillan, and D. L. Dill. Symbolic model checking for sequential circuit verification. *IEEE Trans. on CAD*, 13(4):401–424, 1994.

[BCMD90] J. R. Burch, E. M. Clarke, K. L. McMillan, and D. L. Dill. Sequential circuit verification using symbolic model checking. In *Design Automation Conf.*, pages 46–51, 1990.

[BD05] D. C. Black and J. Donovan. *SystemC: From the Ground Up*. Springer-Verlag New York, Inc., 2005.

[BDFR07] S. Ben-David, D. Fisman, and S. Ruah. Temporal antecedent failure: Refining vacuity. In *International Conference on Concurrency Theory*, pages 492–506, 2007.

[BDM05] B. Becker, R. Drechsler, and P. Molitor. *Technische Informatik — Eine Einführung*. Pearson Education Deutschland, 2005.

[BDTW93] R. R. Bakker, F. Dikker, F. Tempelman, and P. M. Wognum. Diagnosing and solving over-determined constraint satisfaction problems. In *International Joint Conference on Artificial Intelligence*, pages 276–281, 1993.

[Bei90] B. Beizer. *Software Testing Techniques*. John Wiley & Sons, Inc., 1990.

[Ber06] J. Bergeron. *Writing Testbenches Using SystemVerilog*. Springer, 2006.

[BJW04] R. Brinkmann, P. Johannsen, and K. Winkelmann. Application of property checking and underlying techniques. In R. Drechsler, editor, *Advanced Formal Verification*, pages 125–166. Kluwer Academic Publishers, 2004.

[BMA05] B. Bailey, G. Martin, and T. Anderson. *Taxonomies for the Development and Verification of Digitial Systems*. Springer, 2005.

[Bor09] J. Bormann. *Vollständige funktionale Verifikation*. PhD thesis, Technische Universität Kaiserslautern, 2009.

[Bra93] D. Brand. Verification of large synthesized designs. In *Int'l Conf. on CAD*, pages 534–537, 1993.

[BRB90] K. S. Brace, R. L. Rudell, and R. E. Bryant. Efficient implementation of a BDD package. In *Design Automation Conf.*, pages 40–45, 1990.

[Bry86] R. E. Bryant. Graph-based algorithms for Boolean function manipulation. *IEEE Trans. on Comp.*, 35(8):677–691, 1986.

[BS01] J. Bormann and C. Spalinger. Formale Verifikation für Nicht-Formalisten (Formal verification for non-formalists). *Informationstechnik und Technische Informatik*, 43:22–28, 2001.

[BW96] B. Bollig and I. Wegener. Improving the variable ordering of OBDDs is NP-complete. *IEEE Trans. on Comp.*, 45(9):993–1002, 1996.

[CG03] L. Cai and D. Gajski. Transaction level modeling: an overview. In *IEEE/ACM/IFIP International Conference on Hardware/Software Codesign and System Synthesis*, pages 19–24, 2003.

[CGP99] E. M. Clarke, O. Grumberg, and D. Peled. *Model Checking*. MIT Press, 1999.

[CKV01] H. Chockler, O. Kupferman, and M. Y. Vardi. Coverage metrics for temporal logic model checking. In *Tools and algorithms for the construction and analysis of systems*, number 2031 in LNCS, pages 528 – 542, 2001.

[CKV03] H. Chockler, O. Kupferman, and M. Y. Vardi. Coverage metrics for formal verification. In *Correct Hardware Design and Verification Methods*, pages 111–125, 2003.

[Cla06] K. Claessen. A coverage analysis for safety property lists. Presentation at Workshop on Designing Correct Circuits, 2006.

[Cla07] K. Claessen. A coverage analysis for safety property lists. In *Int'l Conf. on Formal Methods in CAD*, pages 139–145, 2007.

[Coo71] S. A. Cook. The complexity of theorem proving procedures. In *3. ACM Symposium on Theory of Computing*, pages 151–158, 1971.

[CS07] H. Chockler and O. Strichman. Easier and more informative vacuity checks. In *ACM & IEEE International Conference on Formal Methods and Models for Codesign*, pages 189–198, 2007.

[DB98] R. Drechsler and B. Becker. *Binary Decision Diagrams – Theory and Implementation*. Kluwer Academic Publishers, 1998.

[DBG96] R. Drechsler, B. Becker, and N. Göckel. A genetic algorithm for variable ordering of OBDDs. *IEE Proceedings*, 143(6):364–368, 1996.

[DCdS07] A. Dias, Jr and D. Cecilio da Silva, Jr. Code-coverage based test vector generation for SystemC designs. In *IEEE Annual Symposium on VLSI*, pages 198–206, 2007.

[DEF+08] R. Drechsler, S. Eggersglüß, G. Fey, A. Glowatz, F. Hapke, J. Schloeffel, and D. Tille. On acceleration of SAT-based ATPG for industrial designs. *IEEE Trans. on CAD*, 27(7):1329–1333, 2008.

[DFGG05] R. Drechsler, G. Fey, C. Genz, and D. Große. SyCE: An integrated environment for system design in SystemC. In *IEEE International Workshop on Rapid System Prototyping*, pages 258–260, 2005.

[DG02] R. Drechsler and D. Große. Reachability analysis for formal verification of SystemC. In *EUROMICRO Symp. on Digital System Design*, pages 337–340, 2002.

[DG05] R. Drechsler and D. Große. System level validation using formal techniques. *IEE Proceedings Computer & Digital Techniques, Special Issue on Embedded Microelectronic Systems: Status and Trends*, 152(3):393–406, May 2005.

[DKBE02] R. Dechter, K. Kask, E. Bin, and R. Emek. Generating random solutions for constraint satisfaction problems. In *Eighteenth national conference on Artificial intelligence*, pages 15–21, 2002.

[DLL62] M. Davis, G. Logeman, and D. Loveland. A machine program for theorem proving. *Comm. of the ACM*, 5:394–397, 1962.

[DP60] M. Davis and H. Putnam. A computing procedure for quantification theory. *Journal of the ACM*, 7:506–521, 1960.

[Dre03] R. Drechsler. Synthesizing checkers for on-line verification of system-on-chip designs. In *IEEE International Symposium on Circuits and Systems*, pages IV:748–IV:751, 2003.

[Dre04] R. Drechsler, editor. *Advanced Formal Verification*. Kluwer Academic Publishers, 2004.

[DS07] S. Disch and C. Scholl. Combinational equivalence checking using incremental SAT solving, output ordering, and resets. In *ASP Design Automation Conf.*, pages 938–943, 2007.

[EEH+06] W. Ecker, V. Esen, M. Hull, T. Steininger, and M. Velten. Requirements and concepts for transaction level assertions. In *Int'l Conf. on Comp. Design*, pages 286–293, 2006.

[EES+07] W. Ecker, V. Esen, T. Steininger, M. Velten, and M. Hull. Interactive presentation: Implementation of a transaction level assertion framework in SystemC. In *Design, Automation and Test in Europe*, pages 894–899, 2007.

[EmV] *EmViD: Embedded Video Detection*. http://www.greensocs.com/GreenBench/EmViD.

[ES04] N. Eén and N. Sörensson. An extensible SAT solver. In *Theory and Applications of Satisfiability Testing*, volume 2919 of *LNCS*, pages 502–518, 2004.

[FD04] G. Fey and R. Drechsler. Improving simulation-based verification by means of formal methods. In *ASP Design Automation Conf.*, pages 640–643, 2004.

[FD06] G. Fey and R. Drechsler. SAT-based calculation of source code coverage for BMC. In *ITG/GI/GMM-Workshop "Methoden und Beschreibungssprachen zur Modellierung und Verifikation von Schaltungen und Systemen"*, pages 163–170, 2006.

[FGC+04] G. Fey, D. Große, T. Cassens, C. Genz, T. Warode, and R. Drechsler. ParSyC: An Efficient SystemC Parser. In *Workshop on Synthesis And System Integration of Mixed Information technologies*, pages 148–154, 2004.

[FKL03] H. Foster, A. Krolnik, and D. Lacey. *Assertion-Based Design*. Kluwer Academic Publishers, 2003.

[FOH93] H. Fujii, G. Ootomo, and C. Hori. Interleaving based variable ordering methods for ordered binary decision diagrams. In *Int'l Conf. on CAD*, pages 38–41, 1993.

[FSBD08] G. Fey, S. Staber, R. Bloem, and R. Drechsler. Automatic fault localization for property checking. *IEEE Trans. on CAD*, 27(6):1138–1149, 2008.

[gco] *gcov – a Test Coverage Program.* `http://gcc.gnu.org/onlinedocs/gcc/Gcov.html`.

[GD03a] D. Große and R. Drechsler. BDD-based verification of scalable designs. In *IEEE International High Level Design Validation and Test Workshop*, pages 123–128, 2003.

[GD03b] D. Große and R. Drechsler. Formal verification of LTL formulas for SystemC designs. In *IEEE International Symposium on Circuits and Systems*, pages V:245–V:248, 2003.

[GD04] D. Große and R. Drechsler. Checkers for SystemC designs. In *ACM & IEEE International Conference on Formal Methods and Models for Codesign*, pages 171–178, 2004.

[GD05a] D. Große and R. Drechsler. Acceleration of SAT-based iterative property checking. In *Correct Hardware Design and Verification Methods*, pages 349–353, 2005.

[GD05b] D. Große and R. Drechsler. *CheckSyC*: An efficient property checker for RTL SystemC designs. In *IEEE International Symposium on Circuits and Systems*, pages 4167–4170, 2005.

[GD06] C. Genz and R. Drechsler. System exploration of SystemC designs. In *IEEE Annual Symposium on VLSI*, pages 335–340, 2006.

[GED07] D. Große, R. Ebendt, and R. Drechsler. Improvements for constraint solving in the SystemC verification library. In *ACM Great Lakes Symposium on VLSI*, pages 493–496, 2007.

[GG07] M. Ganai and A. Gupta. *SAT-Based Scalable Formal Verification Solutions (Series on Integrated Circuits and Systems)*. Springer, 2007.

[GKD05a] D. Große, U. Kühne, and R. Drechsler. Formale Verifikation des Befehlssatzes eines SystemC Mikroprozessors. In *GI Jahrestagung (1)*, volume 67 of *LNI*, pages 308–312, 2005.

[GKD05b] D. Große, U. Kühne, and R. Drechsler. Hw/sw co-verification of embedded systems using bounded model checking. In *IEEE International Workshop on Microprocessor Test and Verification*, pages 133–137, 2005.

[GKD06] D. Große, U. Kühne, and R. Drechsler. Hw/sw co-verification of embedded systems using bounded model checking. In *ACM Great Lakes Symposium on VLSI*, pages 43–48, 2006.

[GKD07] D. Große, U. Kühne, and R. Drechsler. Estimating functional coverage in bounded model checking. In *Design, Automation and Test in Europe*, pages 1176–1181, 2007.

[GKD08] D. Große, U. Kühne, and R. Drechsler. Analyzing functional coverage in bounded model checking. *IEEE Trans. on CAD*, 27(7):1305–1314, 2008.

[GKG+05] D. Große, U. Kühne, C. Genz, F. Schmiedle, B. Becker, R. Drechsler, and P. Molitor. Modellierung eines Mikroprozessors in SystemC. In *ITG/GI/GMM-Workshop "Methoden und Beschreibungssprachen zur Modellierung und Verifikation von Schaltungen und Systemen"*, 2005.

[GLMS02] T. Grötker, S. Liao, G. Martin, and S. Swan. *System Design with SystemC.* Kluwer Academic Publishers, 2002.

[GN03] E. Goldberg and Y. Novikov. Verification of proofs of unsatisfiability for CNF formulas. In *Design, Automation and Test in Europe*, pages 886–891, 2003.

[GPB01] E. I. Goldberg, M. R. Prasad, and R. K. Brayton. Using SAT for combinational equivalence checking. In *Design, Automation and Test in Europe*, pages 114–121, 2001.

[GPKD08] D. Große, H. Peraza, W. Klingauf, and R. Drechsler. Measuring the quality of a SystemC testbench by using code coverage techniques. In E. Villar, editor, *Embedded Systems Specification and Design Languages: Selected contributions from FDL'07*, pages 73–86. Springer, 2008.

[Gro08] D. Große. *Quality-Driven Design and Verification Flow for Digital Systems.* Dissertation, Universität Bremen, Bremen, Germany, October 2008.

[GS97] S. Graf and H. Saidi. Construction of abstract state graphs with PVS. In *Computer Aided Verification*, volume 1254, pages 72–83. Spring, 1997.

[GSD06] D. Große, R. Siegmund, and R. Drechsler. Processor verification. In P. Ienne and R. Leupers, editors, *Customizable Embedded Processors*, pages 281–302. Elsevier, 2006.

[GWDD09] D. Große, R. Wille, G. W. Dueck, and R. Drechsler. Exact multiple control Toffoli network synthesis with SAT techniques. *IEEE Trans. on CAD*, 28(5):703–715, 2009.

[GWKD09] D. Große, R. Wille, U. Kühne, and R. Drechsler. Contradictory antecedent debugging in bounded model checking. In *ACM Great Lakes Symposium on VLSI*, pages 173–176, 2009.

[GWSD08] D. Große, R. Wille, R. Siegmund, and R. Drechsler. Contradiction analysis for constraint-based random simulation. In *Forum on specification and Design Languages*, pages 130–135, 2008.

[GZ03] J. F. Groote and H. Zantema. Resolution and binary decision diagrams cannot simulate each other polynomially. *Discrete Applied Mathematics*, 130(2):157–171, 2003.

[HKHZ99] Y. V. Hoskote, T. Kam, P. Ho, and X. Zhao. Coverage estimation for symbolic model checking. In *Design Automation Conf.*, pages 300–305, 1999.

[HS96] G. Hachtel and F. Somenzi. *Logic Synthesis and Verification Algorithms*. Kluwer Academic Publisher, 1996.

[HT05] A. Habibi and S. Tahar. Design for verification of SystemC transaction level models. In *Design, Automation and Test in Europe*, pages 560–565, 2005.

[Hua05] J. Huang. MUP: A minimal unsatisfiability prover. In *ASP Design Automation Conf.*, pages 432–437, 2005.

[IEE01] IEEE Std. 1346. *IEEE Standard Verilog hardware description language*, 2001.

[IEE05a] IEEE Std. 1666. *IEEE Standard SystemC Language Reference Manual*, 2005.

[IEE05b] IEEE Std. 1800. *IEEE SystemVerilog*, 2005.

[IEE05c] IEEE Std. 1850. *IEEE Standard for Property Specification Language (PSL)*, 2005.

[Int09] Intel. World's first 2-billion transistor microprocessor, 2009. `http://www.intel.com/technology/architecture-silicon/2billion.htm`.

[IPC03] M. K. Iyer, G. Parthasarathy, and K.-T. Cheng. SATORI - a fast sequential SAT engine for circuits. In *Int'l Conf. on CAD*, pages 320–325, 2003.

[IS03] C. Norris Ip and S. Swan. A tutorial introduction on the new SystemC verification standard. White paper, www.systemc.org, 2003.

[Iye03] M. A. Iyer. Race: A word-level ATPG-based constraints solver system for smart random simulation. *Int'l Test Conf.*, pages 299–308, 2003.

[Joh02] P. Johannsen. *Speeding up Hardware Verification by Automated Data Path Scaling*. PhD thesis, Christian-Albrechts-Universität zu Kiel, 2002.

[JPS03] N. Jayakumar, M. Purandare, and F. Somenzi. Dos and don'ts of CTL state coverage estimation. In *Design Automation Conf.*, pages 292–295, 2003.

[KEP06] D. Karlsson, P. Eles, and Z. Peng. Formal verification of SystemC designs using a petri-net based representation. In *Design, Automation and Test in Europe*, pages 1228–1233, 2006.

[KG99] S. Katz and O. Grumberg. Have I written enough properties - a method of comparison between specification and implementation. In *Correct Hardware Design and Verification Methods*, pages 280–297, 1999.

[KGB+06] W. Klingauf, R. Günzel, O. Bringmann, P. Parfuntseu, and M. Burton. Greenbus: a generic interconnect fabric for transaction level modelling. In *Design Automation Conf.*, pages 905–910, 2006.

[KGD07] U. Kühne, D. Große, and R. Drechsler. Improving the quality of bounded model checking by means of coverage estimation. In *IEEE Annual Symposium on VLSI*, pages 165–170, 2007.

[KK07] N. Kitchen and A. Kuehlmann. Stimulus generation for constrainted random simulation. In *Int'l Conf. on CAD*, pages 258–265, 2007.

[Kli05] W. Klingauf. Systematic transaction level modeling of embedded systems with SystemC. In *Design, Automation and Test in Europe*, pages 566–567, 2005.

[KPKG02] A. Kuehlmann, V. Paruthi, F. Krohm, and M. K. Ganai. Robust Boolean reasoning for equivalence checking and functional property verification. *IEEE Trans. on CAD*, 21(12):1377–1394, 2002.

[Kro99] Th. Kropf. *Introduction to Formal Hardware Verification.* Springer, 1999.

[KS05] D. Kroening and N. Sharygina. Formal verification of SystemC by automatic hardware/software partitioning. In *ACM & IEEE International Conference on Formal Methods and Models for Codesign*, pages 101–110, 2005.

[Kue06] U. Kuehne. Modellierung und Verifikation eines RISC Prozessors. Master's thesis, Universität Bremen, 2006.

[KV99] O. Kupferman and M. Y. Vardi. Vacuity detection in temporal model checking. In *Correct Hardware Design and Verification Methods*, pages 82–96, 1999.

[Lar92] T. Larrabee. Test pattern generation using Boolean satisfiability. *IEEE Trans. on CAD*, 11:4–15, 1992.

[LS05] M. H. Liffiton and K. A. Sakallah. On finding all minimally unsatisfiable subformulas. In *Theory and Applications of Satisfiability Testing*, pages 173–186, 2005.

[LTG97] S. Liao, S. Tjiang, and R. Gupta. An efficient implementation of reactivity for modeling hardware in the scenic design environment. In *Design Automation Conf.*, pages 70–75, 1997.

[McM93] K. L. McMillan. *Symbolic Model Checking.* Kluwer Academic Publisher, 1993.

[McM02] K. L. McMillan. Applying SAT methods in unbounded symbolic model checking. In *Computer Aided Verification*, volume 2404 of *LNCS*, pages 250–264, 2002.

[Mic05] ST Microelectronics. TAC: Transaction Accurate Communication. `http://www.greensocs.com/TACPackage`, 2005.

[MLA+05] M. N. Mneimneh, I. Lynce, Z. S. Andraus, J. P. Marques-Silva, and K. A. Sakallah. A branch and bound algorithm for extracting smallest minimal unsatisfiable formulas. In *Theory and Applications of Satisfiability Testing*, pages 467–474, 2005.

[MMMC06] M. Moy, F. Maraninchi, and L. Maillet-Contoz. LusSy: an open tool for the analysis of systems-on-a-chip at the transaction level. *Design Automation for Embedded Systems*, pages 73–104, 2006.

[MMZ+01] M. W. Moskewicz, C. F. Madigan, Y. Zhao, L. Zhang, and S. Malik. Chaff: Engineering an efficient SAT solver. In *Design Automation Conf.*, pages 530–535, 2001.

[MP91] Z. Manna and A. Pnueli. *The Temporal Logic of Reactive and Concurrent Systems: Specification*. Springer-Verlag, 1991.

[MRR03] W. Müller, W. Rosenstiel, and J. Ruf, editors. *SystemC Methodologies and Applications*. Kluwer Academic Publishers, 2003.

[MS99] J. P. Marques-Silva and K. A. Sakallah. GRASP: A search algorithm for propositional satisfiability. *IEEE Trans. on Comp.*, 48(5):506–521, 1999.

[MS05] M. N. Mneimneh and Karem A. Sakallah. Principles of sequential-equivalence verification. *IEEE Design & Test of Comp.*, 22(3):248–257, 2005.

[MSS96] J. P. Marques-Silva and K. A. Sakallah. GRASP – a new search algorithm for satisfiability. In *Int'l Conf. on CAD*, pages 220–227, 1996.

[OMA+04] Y. Oh, M. N. Mneimneh, Z. S. Andraus, K. A. Sakallah, and I. L. Markov. AMUSE: a minimally-unsatisfiable subformula extractor. In *Design Automation Conf.*, pages 518–523, 2004.

[OSC08] OSCI. SystemC, 2008. Available at http://www.systemc.org.

[Par97] T. Parr. *Language Translation using PCCTS and C++: A Reference Guide*. Automata Publishing Co., 1997.

[PBG05] M. R. Prasad, A. Biere, and A. Gupta. A survey of recent advances in SAT-based formal verification. *Software Tools for Technology Transfer*, 7(2):156–173, 2005.

[PH04] D. A. Patterson and J. Hennessy. *Computer Organization and Design*. Morgan Kaufmann Publishers Inc., San Francisco, CA, USA, 2004.

[Pnu77] A. Pnueli. The temporal logic of programs. In *Symposium on Foundations of Computer Science*, pages 46–57, 1977.

[PRB01] T. Petit, J.-C. Régin, and C. Bessière. Specific filtering algorithms for over-constrained problems. *LNCS*, 2239:451–463, 2001.

[PS02] M. Purandare and F. Somenzi. Vacuum cleaning CTL formulae. In *Computer Aided Verification*, volume 2404 of *LNCS*, pages 485–499, 2002.

[RHKR01] J. Ruf, D. W. Hoffmann, T. Kropf, and W. Rosenstiel. Simulation-guided property checking based on multi-valued ar-automata. In *Design, Automation and Test in Europe*, pages 742–748, 2001.

[RS03] J. Rose and S. Swan. *SCV Randomization Version 1.0*, 2003.

[Rud93] R. Rudell. Dynamic variable ordering for ordered binary decision diagrams. In *Int'l Conf. on CAD*, pages 42–47, 1993.

[SDGC07] J. Simmonds, J. Davies, A. Gurfinkel, and M. Chechik. Exploiting resolution proofs to speed up LTL vacuity detection for BMC. In *Int'l Conf. on Formal Methods in CAD*, pages 3–12, 2007.

[Sem03] Semiconductor Industry Association. *International Technology Roadmap for Semiconductors*, 2003.

[Sem06] Semiconductor Industry Association. *International Technology Roadmap for Semiconductors*, 2006.

[Sht01] O. Shtrichman. Pruning techniques for the SAT-based bounded model checking problem. In *Correct Hardware Design and Verification Methods*, volume 2144 of *LNCS*, pages 58–70, 2001.

[SKF+09] A. Sülflow, U. Kühne, G. Fey, D. Große, and R. Drechsler. WoLFram - a word level framework for formal verification. In *IEEE/IFIP International Symposium on Rapid System Prototyping (RSP)*, pages 11–17, 2009.

[SKW+07] A. Sülflow, U. Kühne, R. Wille, D. Große, and R. Drechsler. Evaluation of SAT like proof techniques for formal verification of word level circuits. In *IEEE Workshop on RTL and High Level Testing*, pages 31–36, 2007.

[SMS+07] T. Sakunkonchak, T. Matsumoto, H. Saito, S. Komatsu, and M. Fujita. Equivalence checking in C-based system-level design by sequentializing concurrent behaviors. In *International Conference on Advances in Computer Science and Technology*, pages 36–42, 2007.

[Sny08] W. Snyder. Dinotrace, 2008. Available at `http://www.veripool.org/wiki/dinotrace`.

[Som98] F. Somenzi. *CUDD: CU Decision Diagram Package Release 2.3.0*. University of Colorado at Boulder, 1998.

[Som01] F. Somenzi. Efficient manipulation of decision diagrams. *Software Tools for Technology Transfer*, 3(2):171–181, 2001.

[SSS00] M. Sheeran, S. Singh, and G. Stålmarck. Checking safety properties using induction and a SAT-solver. In *Int'l Conf. on Formal Methods in CAD*, volume 1954 of *LNCS*, pages 108–125. Springer, 2000.

[SVV04] A. Smith, A. Veneris, and A. Viglas. Design diagnosis using Boolean satisfiability. In *ASP Design Automation Conf.*, pages 218–223, 2004.

[Sys02] *SystemC 2.0 user's guide*. http://www.systemc.org, 2002.

[Sys03] SystemC Verification Working Group, http://www.systemc.org. *SystemC Verification Standard Specification Version 1.0e*, 2003.

[TK01] S. Tasiran and K. Keutzer. Coverage metrics for functional validation of hardware designs. *IEEE Design and Test of Computers*, 18(4):36–45, 2001.

[Tse68] G. Tseitin. On the complexity of derivation in propositional calculus. In *Studies in Constructive Mathematics and Mathematical Logic, Part 2*, pages 115–125, 1968. (Reprinted in: J. Siekmann, G. Wrightson (Ed.), Automation of Reasoning, Vol. 2, Springer, Berlin, 1983, pp. 466–483.).

[TVKS08] D. Tabakov, M.Y. Vardi, G. Kamhi, and E. Singerman. A temporal language for SystemC. In *Int'l Conf. on Formal Methods in CAD*, pages 1–9, 2008.

[Var07] M. Y. Vardi. Formal techniques for SystemC verification. In *Design Automation Conf.*, pages 188–192, 2007.

[Vel04] M. N. Velev. Efficient translation of boolean formulas to CNF in formal verification of microprocessors. In *ASP Design Automation Conf.*, pages 310–315, 2004.

[Weg87] I. Wegener. *The Complexity of Boolean Functions*. John Wiley & Sons Ltd., and B.G. Teubner, Stuttgart, 1987.

[WFG+07] R. Wille, G. Fey, D. Große, S. Eggersglüß, and R. Drechsler. Sword: A SAT like prover using word level information. In *VLSI of System-on-Chip*, pages 88–93, 2007.

[WGHD09] R. Wille, D. Große, F. Haedicke, and R. Drechsler. SMT-based stimuli generation in the SystemC verification library. In *Forum on specification and Design Languages*, 2009.

[WTSF04] K. Winkelmann, H.-J. Trylus, D. Stoffel, and G. Fey. Cost-efficient block verification for a UMTS up-link chip-rate coprocessor. In *Design, Automation and Test in Europe*, volume 1, pages 162–167, 2004.

[YAPA04] J. Yuan, A. Aziz, C. Pixley, and K. Albin. Simplifying boolean constraint solving for random simulation-vector generation. *IEEE Trans. on CAD*, 23(3):412–420, 2004.

[YPA06] J. Yuan, C. Pixley, and A. Aziz. *Constraint-based Verification*. Springer, 2006.

[YSP+99] J. Yuan, K. Shultz, C. Pixley, H. Miller, and A. Aziz. Modeling design constraints and biasing in simulation using BDDs. In *Int'l Conf. on CAD*, pages 584–590, 1999.

[ZM03] L. Zhang and S. Malik. Validating SAT solvers using an independent resolution-based checker: Practical implementations and other applications. In *Design, Automation and Test in Europe*, pages 880–885, 2003.

[ZMMM01] L. Zhang, C. F. Madigan, M. H. Moskewicz, and S. Malik. Efficient conflict driven learning in a Boolean satisfiability solver. In *Int'l Conf. on CAD*, pages 279–285, 2001.

[ZSM+05] J. Zhang, S. Sinha, A. Mishchenko, R. Brayton, and M. Chrzanowska-Jeske. Simulation and satisfiability in logic synthesis. In *Int'l Workshop on Logic Synth.*, pages 161–168, 2005.

Index

FSC
www.fsc.org
MIX
Papier | Fördert
gute Waldnutzung
FSC® C083411

Kolibri 360 GmbH
Ferdinand-Jühlke-Straße 7
99095 Erfurt, Deutschland
produktsicherheit@kolibri360.de